THE TWIN DIMENSIONS

The TWIN DIMENSIONS

INVENTING TIME AND SPACE

Géza Szamosi

McGraw-Hill Book Company

New York St. Louis San Francisco
Toronto Hamburg Mexico

Figure 4-2 (Zeus [Prometheus]) courtesy of Republic of Greece, Ministry of Culture.
Figure 4-3 (Egyptian Boatmen) from JANSON, page 44.
Figure 4-4 (Aphrodite by Praxiteles) courtesy of Monumenti Museie Gallerie Pontificie, Vatican.
Figure 6-1 (Albrecht Dürer, *Draftsman Drawing a Reclining Nude*) from K. A. Knappe, *The Complete Engravings, Etchings and Woodcuts* [New York: Abrams], page 373.
Figure 9-4 (Pablo Picasso, *Portrait of Daniel Henry Kahnweiler*) courtesy of The Art Institute of Chicago.
Figure 9-5 (René Magritte, *The False Mirror*) courtesy of the Museum of Modern Art, New York.
Figure 9-7 (Jackson Pollock, *Cathedral*) courtesy of Dallas Museum of Art.
The following figures are from *The McGraw-Hill Encyclopedia of Art,* all plates of which were copyrighted © 1958 by the Institute per la Collaborazione Culturale, Rome, under the Universal Copyright Convention: 4-1 (cube statue of Saqqara: Egyptian Museum, Cairo [IV, pl. 354]); 4-5 (*The Victory of Samothrace:* Louvre, Paris [VII, pl. 177]); 6-2 [XI, pl. 91]; 6-3 (Donatello, *The Feast of Herod:* Baptistery, Siena [IV, pl. 242]); 6-4 (Ideal city: Galleria Nazionale delle Marche, Urbino [XI, pl. 88]); 6-5 (Raphael, *The School of Athens:* Vatican [XI, pl. 426]); 9-1 (Velasquez, *View of the Garden of Villa Medici in Rome:* Prado, Madrid [XIV, pl. 323]); 9-2 (Monet, *Water-lily Pool:* Louvre, Paris [X, pl. 130]); 9-3 (Braque, *The Portuguese:* Kunstmuseum, Basel [II, pl. 348]); and 9-6 (Mondrian, *Composition in Red, Yellow, and Blue:* Gemeentemuseum, The Hague [X, pl. 123]).

All quotations, as credited in the Notes and Bibliography, are by permission, unless best efforts to find copyright holder were unsuccessful.

1 2 3 4 5 6 7 8 9 DOCDOC 8 7 6

ISBN 0-07-062646-4

LIBRARY OF CONGRESS CATALOGING-IN-PUBLICATION DATA

Szamosi, Géza.
The twin dimensions.
Bibliography: p.
1. Space and time—Popular works. I. Title.
QC173.59.S65S99 1986 530.1'1 85-25610
ISBN 0-07-062646-4

BOOK DESIGN BY PATRICE FODERO

Contents

THE TWIN DIMENSIONS

1 Introduction

Of the millions of creatures living on earth, human beings are the only ones who want to know what lies beyond their immediate environment; they are the only ones who care what happened before they were born and speculate on what will happen after they are gone. In other words, they are the only creatures interested in space and time.

This book is about time and space and about the human interest in it. Since I am a physicist with a lifelong interest in the subject, it is no surprise that this book contains chapters about the large-scale time and space of contemporary cosmology as well as about time and space at the other end of the cosmic scale, in the less often described though no less distant worlds of the elementary particles. Originally I intended the book to be about this and nothing else.

But as I started to write, I began to come across other aspects of my subject matter which did not belong to either physics or to astronomy. Problems concerned not so much with space and time "out there" as with the space and time of our mind. How and where, for example, did the human brain, which is so small in volume and has so brief an existence, become equipped to accommodate and to handle enormous extensions of time and space? How and where did this ability start? How and when did human beings actually "discover" that there *is* time and space? How did our brains evolve to perceive and organize the world always in the framework of time and space? And how did we explore these twin dimensions throughout our history?

In trying to find answers to such questions, I not only ended up with more questions, but I also learned a lot of fascinating things, and this enabled me to enlarge the scope of the book. Although still concerned with the time and space of physics and astronomy, it also contains much of what I learned about time and space in other contexts: in the contexts of biological evolution, of the workings of the brain, and of the sociocultural history of our species. The book became a sketch of how space and time was "invented" in the perceptions of early life-forms and evolved to the space and time of contemporary civilization.

To try to pull all this together, I will first present a bird's-eye view of where we are heading and by what route. Our journey starts in the next chapter with biology; how life, in the process of adapting to the environment, found out about properties of space and time.

I became involved with the biological aspects of time and space when I began to wonder how the human mind first "discovered" time and space. I soon learned, however, that long before the appearance of homo sapiens, other life-forms had already explored time and space and our species has inherited much of what our more primitive ancestors had learned. Thus, biological evolution has equipped us with an innate mastery of space and time which is still, in many ways, superior to anything we can learn from physics.

Ask an average, intelligent person with a normal educational background but no special training in physics the following question: A ball is thrown against a wall from a certain distance with a certain speed at a certain angle. Where will the ball be after so and so many seconds? As a rule, such a person will not be able to calculate the answer. Without a pretty fair knowledge of the laws of mechanics, he or she cannot find the correct answer. But put a ball in the hand of the very same person and ask them to bounce it off a wall and catch it, and they would have no trouble at all. Just from perceiving the initial speed, angle, and distance (the data given in the example), they will know exactly where the ball will be and when (the question asked) and will even be able to take into account the effects of gravity and air resistance. The solution of a competent player is every bit as good as the solution of a competent physicist. Both obtain from the same given conditions the same correct answer—except that the player does it much faster than the physicist.

In more complex situations, the competition becomes hopeless—for the physicist. Adding insult to injury, not only can any healthy human brain solve practical space and time puzzles faster and better than a trained physicist, but even an animal can do so. Imagine the problems a fox faces when it pursues a rabbit, or the rabbit faces when it tries to escape the fox. The brains of both animals receive information through their respective senses about each other's speeds, positions, directions, and accelerations, as well as about some often very complicated features of the environment, such as obstacles and alternate routes, together with an enormous amount of irrelevant data. Their brains suppress the irrelevancies, select and process the relevant data, "compute" each other's instant speeds and positions, make predictions if necessary, send signals to the motoric system to perform some often highly complex motion appropriate for the situation, and all this in a split second.

The whole process is so intricate and demands such extraordinary capacity to compute rapidly changing patterns in space and time that not only a single physicist but a whole think tank with a battery of computers would be pathetically inadequate to the task. Saying, as we often do, that the animals do this "instinctively" is merely giving a name to a process that we do not yet understand. Giving it a name does not, of course, explain it. The fact is that many living creatures, humans included, have a prewired, innate capacity to make immediate sense of complicated patterns in space and time that are, as yet, much too complex for a conscious intellectual process. This fact suggested to me that one useful route to take in an inquiry about time and space may start with biological evolution.

The most powerful and versatile tool that evolved to aid in the process of adaptation is the brain. The third chapter is devoted to some of its history, following the ideas of Harry Jerison of the University of California at Los Angeles. Jerison has suggested a theory to describe how and why the mammalian brain—of which the human is the most developed variant—evolved to have the property of perceiving the external world in the framework of space and time. The theory starts with the recognition that the way a living creature experiences the external world depends on two factors: on the information the creature's senses can detect in the environment and on the way the nervous system processes that information. Jerison's theory describes how during the last quarter of a billion years the mammalian nervous system, in adapting to the environment, evolved to analyze and organize incoming sensory information in such a way that the world it experienced became a world of separate and permanent objects in extended space and time. Mammals, and therefore humans, see and can make sense of the world in this framework only. The world of permanent objects in extended space and time is their worldview, determined by biological evolution; this is the only world they can ever experience; this, in short, is the mammalian cosmology.* And this is also as far as biology alone can take us in our search for the meaning of space and time.

A new and uniquely human world of time and space started with the evolution of language. A couple of simple examples will make this clear. An intelligent animal, like a dog, can understand where its food is located. It can also be trained to expect from signals that its food will soon be at a certain place. But no dog can ever be trained to understand that "the food was there yesterday" or that "it will be at that other place tomorrow." A cat can be trained to go to a certain place and it often will go voluntarily to explore new places. But it can never be made to understand the existence of a place it has not seen or felt. No animal brain has the

* I shall use the word "cosmology" not only in its technical sense, meaning the science of the universe, but also metaphorically to characterize the overall experience of the world by an organism or the "worldview" of a particular human civilization. It is in this sense that one speaks about a "reptilian cosmology" or the "cosmology of early medieval Europe."

capacity to understand any kind of reference to times before its birth or to times after its extinction or to places it has never been. The animal world of time and space is therefore restricted.

In contrast, the evolution of language has allowed our mental world of space and time to become limitless. But this world is not perceivable. It is purely symbolic. When I describe where I was an hour or a year ago, when Homer wrote about the Trojan War, or when Einstein calculated the motion of the perihelion of the planet Mercury, it was all done by using symbols: words, numbers, and the like. When we measure the length of an object or estimate some time span, the results are expressed in units and numbers which are, again, human symbols. Thus we can speak of *symbolic time and space* accessible to the human mind only.

There are a great variety of symbolic times and spaces in human culture. Not all are characterized by words or numbers. Paintings, statues, buildings, maps, holy places, eternal hunting grounds, and other netherworlds are all examples of symbolic spaces. Rhythms, melodies, stories, plays in a theater, poetic meters, holy days, and eternity, all signify symbolic times. These are all different from each other, and all are different from the, say, symbolic time a digital clock describes with digits. None of these symbolic dimensions makes any sense to any animal. But when humans refer to "space and time," they usually mean symbolic space and time in one of its many forms. The second half of Chapter 3 contains some speculations about the evolutionary origins of human symbolic time and space.

A symbol is just a symbol and never the real thing. The word "water" is not wet and it does not flow. Yet symbols have great power and can make a tremendous impact on the real world. The appearance of human symbolism was one of the greatest evolutionary events in the whole history of life. It has already changed much of the environment on earth and it is on its way to making its mark on the solar system and perhaps beyond. Chapter 4 deals with some early symbolic concepts of space and time.

Since symbols are created in specific human cultures, different civilizations evolved different symbolic cosmologies and thus perceived and described the world in the framework of different symbolic times and spaces. An ancient Egyptian myth is based on a time-image which is as different from the one John Milton used in *Paradise Lost* in the seventeenth century as that is different from what we find in James Joyce's twentieth-century novel *Ulysses.* But not only the abstract concepts are different. The very perceptions of time and space, the way people see and feel the world, also evolved differently in different civilizations. Compare, for example, space as represented in a Babylonian relief with the space we see in a painting by the Dutch contemporary of Milton, Jan Vermeer. The two are as different from each other as both are from the space depicted by a twentieth-century artist.* The history of human symbolic times and spaces is likely to be as rich as it is little known. However, except for the discussions of biology and the excursion, in Chapter 4, into some prescientific ideas of time and space, we shall deal here exclusively with the symbolic times and spaces of Western civilization. It is, to put it mildly, a large enough topic.

An important new kind of symbolic time and space, and with it a new human cosmology, evolved in the seventeenth century with the birth of the experimental sciences: the scientific revolution. The first experimental sciences described the world in time and space with mathematical, numerical symbols and assumed that these numbers could be found by *careful experimentation and observation.* The "experimental method," as it came to be known, was based on the idea that the human senses, used judiciously, are capable of obtaining reliable information from the external world and that language and mathematics can be used to formulate general laws in symbolic time and space. This was an extraordinary idea, the like of which had never been consistently tried before.

The idea of using the senses to find numerical law and order in time and space gave birth to the sciences.† But the idea itself originated not from the sciences but rather in the arts. About four centuries before the birth of the experimental method in physics, polyphonic music evolved in Western Europe and brought with it

* In contrast, human beings' inborn biological mastery of time and space is and probably has always been pretty much the same all over the world.

† To avoid misunderstanding, the word "science" in this book *always* denotes the natural sciences only, although there are no satisfactory definitions separating natural and human and social sciences. "Science" in this book means, roughly speaking, the domains of physics, astronomy, chemistry, and fundamental biology and the areas where these domains overlap.

the notational system for measured musical rhythms. This system made possible the first number-based and environment-independent accurate symbolic time measurements in history. Its success showed convincingly that it was possible to use numbers to describe the passing of time by measuring short time intervals in a precise and reliable way. Therefore, long before metric or mathematical time had ever been used in science, it had already been invented, defined, used, and studied by musicians. In Chapter 5, I describe music's role in the evolution of the modern concept of time.*

The visual arts, on the other hand, were instrumental in the discovery of laws to describe the perception of space. A couple of centuries before the experimental method was established in science, sophisticated numerical rules were found for the most important human spatial sense: vision. The artists of the Italian Renaissance applied the laws of geometry to the laws of vision, which allowed for the first time in history the creation of highly realistic pictures. This result suggested strongly that human vision may well obey exact mathematical laws in its perception of space and spatial properties. A sketch of this development is related in Chapter 6.

One of the few things on which cultural historians of Western civilization all seem to agree is that three important developments occurred, took root, and flourished in Western civilization only. In many ways they characterize our civilization: polyphonic music, perspective painting, and experimental science. It is remarkable, though seldom noticed, that all three came into existence as people struggled with basically the same technical problem: how to use the senses to find reliable measures of time intervals, spatial distances, and their various relations. How to impose, in other words, mathematical order on the perceivable world. It is also remarkable—and equally seldom noticed—that in this crucial enterprise, the arts preceded experimental science.

Successful solutions to these problems in Western art and science were found slowly but progressively from approximately the

* The "experimental method" was not invented in any one year, of course. For simplicity's sake, I shall refer to the year 1600 as the date of the beginning of experimental science. This date is round enough and corresponds approximately to the date of Galileo's first mechanical experiments.

mid-thirteenth century (the beginnings of polyphony) to the time of Isaac Newton (the end of the scientific revolution) in the late seventeenth. These new developments gave birth to a new and powerful system of symbolic time and space which amounted to a new way of perceiving and understanding the world. As I mentioned above, this new cosmology evolved as people consciously started to trust their senses and direct sensory impressions became the main source of information about the nature of the external world. This happened during this period for the first and only time in known history. In all other human explanations of the world, in all other symbolic cosmologies, the role of sensory impressions was always secondary, and whatever factors were imagined to be important in the workings of the external world, the need to expose these to the controls of the senses never arose.

Early modern European civilization broke with this very long tradition, and this had important consequences. New, sophisticated notions were introduced which reflected what the senses perceived. As a result, a new view of the world, a new symbolic cosmology, evolved which gradually became compatible with our inborn picture of the external world: the mammalian cosmology of the human brain. The world was now seen as well as thought of as consisting of individual and permanent objects moving in extended space and time. Instead of being governed by gods residing in mythological-religious netherworlds, the workings of the world were now determined by causal laws governing predictable and perceivable processes both in nature and in society. It was in this new mental framework of space and time that classical physics was formulated, European art created and perceived, and Western history and literature, philosophy, music, and mathematics evolved.

I shall refer to this whole framework of symbolic time and space as "classical." This name fits not only because this framework is associated with classical physics and classical art, but also, as discussed in Chapter 4, because some of its ideas originated in antiquity; in the spatial concepts of the ancient Greeks and in the time notions of the Hebrew Bible. The classical space and time of Europe is the subject of Chapter 6.

The rest of the book is devoted to the symbolic time and space of our own century. Although we still experience the directly observable world through a mammalian cosmology, our symbolic cosmology has, during the past eighty years, broken out of the classical framework both in the arts and sciences. It was at the beginning of the twentieth century that a new symbolic time and space was invented and with it a new human cosmology. The emergence of this new mental framework was a historic event of the first magnitude.

While the evolution of the mental framework of classical time and space was slow, its replacement was swift. Instead of taking centuries to evolve, the new view of the world, the "modern" notions of time and space, emerged within the first decade of this century almost simultaneously from works in physics, in painting, and in music.

We see now in hindsight that around the beginning of this century physics reached a stage where the world of human dimensions, the world known by direct sensory impressions, became more or less understood. Interest then shifted toward deeper layers, toward phenomena which take place in the worlds of the very small or the very large or those that take place very fast. About these worlds, we can have only secondhand or indirect sensory information. The information coming from these extreme dimensions contained shocking news. Most phenomena in these worlds could not be understood in the mental framework of classical space and time, and they were often totally alien to the inborn intuitive cosmology of our brain.

Investigating the very fast phenomenon of the propagation of light in 1905, Albert Einstein created the special theory of relativity. This theory made many of the basic concepts of classical time and space obsolete and replaced it with a new symbolic framework of time and space. Einstein's work marked the beginning of a long period of epoch-making physical and astronomical discoveries all relating to the framework of time and space. The special theory of relativity was followed by the even more profound general theory ten years later. Another ten years saw the discovery of the expansion of the space of the universe and with it the possibility of directly observing events of the distant past. Every decade of this century produced at least one important new idea relating to time and space, and often these were the most important discoveries of the period. The present is no exception. Much of the fundamental research in the 1980s is concerned with hidden dimensions of space and how these influence the workings of the universe.

The first decade of this century also saw a successful attack on other elements of the inborn cosmology of our brain. In 1900, Max Planck discovered that the continuity we think we perceive in the external world does not exist in the fundamental processes of nature. In 1905, Einstein produced a hypothesis concerning the nature of light which not only contradicted classical physics but was completely alien to every earlier experience concerning the nature of the world. These discoveries led, in another twenty years, to the evolution of quantum mechanics and a totally new scientific worldview not only incompatible with all earlier ideas but also clashing head-on with our direct sensory experiences, our intuitive, inborn mammalian cosmology. Chapters 7, 8, and 10 are about these developments.

The visual arts underwent a similar revolution. Painters of the early twentieth century were also preoccupied with problems of space and visual appearances. They felt that the then accepted ways of depicting space and spatial forms were inadequate in accommodating their new ideas. This feeling resulted in a burst of activity after 1907, the year that a group of pioneering artists, with Pablo Picasso and Georges Braque in the lead, broke radically with the old ways of painting. These artists depicted visual appearances in wholly new ways and created sights, the likes of which had never been seen before. Their method and style became known as "cubist."

This was the greatest revolution in painting since the Renaissance, and it was also a first step in a whole series of radical innovations and discoveries. A host of other schools of painting followed cubism. Their influence soon spread to all forms of the plastic arts. Virtually every decade of the twentieth century offered new ideas and methods in all the visual arts. Most had little in common with cubism except in the break with the classical ideas about the perception of space and spatial forms. Cubism was, however, the detonator of the process which replaced the classical symbolic space of paintings with a host of other symbolic spaces, and introduced wholly new notions about the nature of human vision and the visible world. Chapter 9 contains some details about these developments.

As for the sense of time, it was also challenged and transformed—by the changes which took place in Western music. As in physics and painting, the changes in the twentieth century were radical, perhaps the most radical in the history of music. The pivotal event here was the introduction of what became known as "atonality." The originator of this daring innovation was Arnold Schoenberg and the date was 1908. In spite of strong (and still by no means conquered) opposition to the idea, atonality became the source of most of the new ideas in twentieth-century musical theory and practice. In Chapter 9, I will argue that atonal music corresponds to a time sense which is different from and incompatible with those of virtually all previous modes of music. Therefore atonal music represents a new type of symbolic time.

Twentieth-century civilization has been very much under the influence of these cultural revolutions of the early years of the 1900s. The new mental framework of symbolic time and space opened up a new world with new challenges and opportunities. Literature is a good example: Already in the first two decades of the century, writers like Marcel Proust, James Joyce, and Franz Kafka made use of images of time and space in the human mind which would have been unimaginable before our century was born. Thus, a new, modern symbolic framework of time and space made the forms of all twentieth-century art and science radically different from any that went before. It was the source of much of what came to be called "modernism" by cultural historians.

The last chapter takes us to the present and to a glimpse into ongoing research in physics and astronomy. A new insight emerging from current research is that the properties of elementary particles may contain very important clues to an understanding of the large-scale properties of the universe. This is an astonishing idea and one that is likely to have extraordinary consequences. Currently, large particle accelerators are being built in several countries with the aim of studying the behavior of matter in very small time and space intervals. These accelerators are serving, in some sense, as super microscopes allowing for the observation of processes taking place deep inside very small regions of the world. It is currently thought that these same accelerators could also be used for studying the state of the universe at its very beginnings. Like magic mirrors in a particularly imaginative mythology, these instruments may allow us to probe the primordial fabric from which our universe in space and time evolved about fifteen billion years ago and to find out how this process took place. But here we are not indulging in mythology, nor even in philosophy. These ideas were inspired by experiments and observations; they can lead to predictions which can again be tested by new experiments and new observations. After the many enchanting and colorful mythologies, we are getting closer to the real thing this time.

The worlds of the very small and the very large and how they were once linked in the history of the universe are then the subjects of the last chapter of this book. Our starting point, on the other hand, is a phenomenon which may seem more modest but is in fact no less impressive. We now turn to the origins of all our proud achievements, to life's first adaptive recognition of space and time.

2

Life Adapts to Time and Space

(The Discovery of Patterns)

For now hath time made me his numbering clock
Shakespeare, *Richard II*

All living things on our planet experience environmental changes in cyclic, periodic patterns. These cyclical patterns are created by the internal mechanics of the solar system. All components of the solar system move periodically; i.e., in recurring cycles. They either rotate around an axis, orbit around a center, or, most often, do both. To creatures on earth, these processes manifest themselves in never-ending orderly sequences. Day follows night follows day forever, the moon goes through its monthly cycle again and again, the seasons repeat their rhythms, ebb and tide follow one another in unending succession. These changes have a powerful effect on the environment. Consequently, all living things experience them, and in order to survive they must adapt to them.

The strategy of adapting to these patterns is essentially the same in all life-forms. It consists of the evolution of some internal physiological mechanism to regulate the temporal behavior of the organism. These mechanisms are usually called "biological" or "internal" clocks and they are present in virtually all forms of life and at every level of organization. All biological clocks, no matter how different their structure or the way they work, have one function in common: to synchronize an organism's behavioral patterns with the external environment. Present probably since the beginning of life on earth, biological clocks were thus the first organizers of time in life—the first to give life a temporal structure.

For most living things the most powerful rhythm in nature is the twenty-four-hour cycle of day and night. Not surprisingly, this period governs a great variety of behavioral time patterns in living creatures from protozoa to the human species. To see how some of these processes work, let us take a look at some examples.

The paramecium is a type of protozoan that exchanges genetic material by conjugation. In this process, two cells join and inseminate each other. Conjugation occurs once a day and, under normal circumstances, at daytime only. But surprisingly, it is not sunlight or the warmer daylight temperature which triggers the process. If the paramecia are kept constantly in darkness and at nighttime temperatures, they will still mate once a day and during the day, sun or no sun, warmth or no warmth. The obvious conclusion is that the paramecia have internal clocks which regulate their behavior quite independently of the environment.

We have all experienced our biological clocks subjectively. But they have also been investigated more objectively in relatively simple experiments with human volunteers. In one type of experiment, the subjects were kept in a controlled environment, in deep caves or in buildings, in complete isolation. All external cues to daily rhythms were removed. The volunteers did not know whether it was day or night, how long they had slept, or how much time had elapsed. Temperature, humidity, and illumination were all kept constant. These human guinea pigs were isolated from the sights, sounds, and smells which, in normal social surroundings, would have signaled the rhythms of "social time." In all cases the subjects of such experiments maintained their day-night rhythms. They slept, were awake, got hungry, ate, emptied, in short, performed their bodily functions, in an approximately twenty-five to twenty-seven-hour cycle. Some deviation from the exact twenty-four-hour periodicity was the rule, but these deviations were seldom larger than about 10 to 15 percent.

Human beings have a large number of internal chronometers. Virtually every human physiological factor which has been investigated for daily periodicity has been shown to possess it. These internal chronometers manifest themselves in many ways. There are rhythmic changes in body temperature, in endocrine chemistry, in nervous activities, and in metabolism. There is evidence of a daily rhythm in human (and animal) sensitivities to drugs, poisons, biological agents, and allergens. A large number of daily rhythms have been observed in the individual cells of our various organs. A fast-growing medical discipline has recently evolved which attempts to utilize these rhythmic responses of the human body for therapeutic purposes. It is truly amazing that such a large number of timekeeping mechanisms could all work in phase, helping, instead of canceling, each other out. That they do this implies an extraordinary degree of order in our whole timekeeping organization.

The daily rhythms of organisms are often called the "circadian" rhythms. The word derives from the Latin *circa* (about) and *dies* (day). The reason for the compound name is that the period of the circadian rhythms is twenty-four hours in the presence of such "environmental cues" as sunrise and sunset but, in the absence of such cues, is only approximately twenty-four hours.

There are also noncircadian biological clocks which govern other temporal patterns. In fact, with the exception of the approximately eleven-year cycle in the sun's activity, all periodic events which influence the earth's environment have corresponding biological clocks with matching periods.* A famous example is the clock which governs the extraordinary pattern of the spawning runs of the grunion.

The grunion is an ordinary small fish living in the coastal waters of California; its only claim to fame is that evolution enabled it to synchronize its spawning habits with the finer workings of the tide. At highest, or "spring" tide, either at full moon or at new moon, during their periods of reproduction, the fish are able

* The reverse would not be true. The timekeeping mechanisms which regulate, for example, the periods of our heartbeats, our breathing, or the discharges of our neurons evolved in response to the needs of our internal environment. There are no recurring processes with matching periods in the external world.

to reach the highest water line on the beach, where they fertilize and bury their eggs. They do this in the brief time between two waves. For the next two weeks, the eggs incubate in safety, out of reach of waves even at the high tide. Two weeks is the necessary incubation period and is, of course, the length of time before the next spring tide occurs. The eggs hatch, and the young fish are washed back into the sea. The exact timing of the breeding of the grunion is particularly striking because the spawning pattern has become synchronized, not just with the period of the daily tide and ebb, but with the period of the highest tide, which is fourteen and seven-tenths days.

But the ability to anticipate and exploit periodic events is widespread. A large number of coastal marine organisms, for example, synchronize their behavior with the tides—which are, on the average, just under twelve and a half hours apart—and this behavior persists when the creatures are kept in a laboratory, far away from the influence of the tides.

Annual hibernation is also a periodic activity. Experiments with hibernating ground squirrels have shown that their yearly hibernating pattern persists for over three years even if their habitat is kept at constant temperature, they have ample supplies of food and water, and artificial light eliminates the shortening of days. It seems then that there exist internal clocks whose period is approximately a year. In this they resemble calendars, but they nevertheless work like clocks.

Most observations lead to the conclusion that biological rhythms are inborn. The clocks which trigger an organism's rhythmic behavior are genetically programmed to be tuned always to the same time period. These periods evolve in adaptation to the environment, to be sure, but once established, the internal clocks work autonomously. They are triggered not by environmental events, but by genetic memory. If this assumption is true, then biological clocks should work on the same period even on a spaceship traveling in regions far removed from the geophysical happenings of the earth. As yet, such observations have not been performed on board spaceships, but a striking experiment in 1962 by K. C. Hamner of the University of California at Los Angeles came pretty close. Hamner and his co-workers took hamsters, fruit flies, cockroaches, bean plants, and bread molds to the South Pole. All these organisms were known to exhibit circadian rhythms in normal environments—the hamsters in their running habits, the fruit flies in their emergence from pupae, the bean plants in leaf movements, and the bread mold in growth. At the South Pole, they were placed on a turntable which rotated at the same speed as the earth but in the opposite direction. It was assumed that all effects of the earth's rotation—day-night pattern, magnetic field changes, periodic changes in humidity, in air pressure, and in temperature—had thereby been eliminated. In such circumstances, it would seem that the time patterns of the solar system would be suspended, that there would be no earthly indications of the daily time. Yet the circadian rhythms of the activities of all the organisms in the experiment continued unchanged. In addition, all their activities remained the same, whether the turntable was rotated or not.

Since the *periods* of the biological clocks evolved genetically, these cannot be changed. But the *phase* of these clocks can be altered. We experience just such alterations, called entrainments, after long-distance east-west or west-east flights. We find upon arrival that the daily routine of life is out of phase with our new environment. After some time, the new environment entrains our biological clocks toward the right phase. Our clocks still observe the day-night rhythm, and their periods remain unchanged, but with a new phase corresponding now to the actual phase of the environment.

The clocks in all creatures can be entrained in a similar way if we expose them, with the help of artificial light, to any day-night pattern with a twenty-four-hour frequency. The organisms' biological clocks will, after some time, work in accordance with the new, artificial rhythm of the environment.

It seems that most biological clocks are controlled by biochemical processes. Among vertebrate animals, the pineal gland, a small organ in the brain, seems to be much involved in general timekeeping. But the details of the enormously complex processes associated with either the workings of individual clock mechanisms or the synchronization of a large number of such clocks are not really understood as yet. The individual clocks probably work on a molecular level and must be exquisite constructions. The behavior of one marine creature, for example, the one-celled alga called *Gonyalna polyedra,* is controlled by no less than four different clocks, each triggering periodic changes once a day. The four mechanisms act at different times during the twenty-four-hour period and govern the onset of such diverse activities as photosynthesis, cell division, luminescence, and flashing. And all this machinery, all the timekeeping and instructions for action, are in a single microscopic cell.

All biological clocks have to be able to communicate information to the relevant parts of an organism, they have to be able to absorb fuel to power themselves, and they have to have a mechanism for resetting themselves so that after the required time they are able to work again. The clocks also possess some mechanism to take care of entrainment, and, in addition to all this, they are sturdy enough to withstand normal changes of temperature, humidity, air or water pressure, and so forth. To evolve such complex machinery was a tremendous task and nothing shows the importance of biological clocks better than the fact that they evolved at all.*

While night inevitably follows day, their relative lengths do change slowly throughout the year, offering another predictable time pattern. Evolution did not miss the opportunity to make use of such nuances. There are biological clocks which, instead of working on fixed geophysical periods, are tuned to the subtle changing of the relative lengths of day and night. The behavior governed by such clocks is called photoperiodism. This phenomenon was discovered in 1920 by two American botanists, W. W. Gardner and H. A. Allard. They observed that the flowering of plants is influenced by the relative lengths of day and night. This was a wholly unexpected discovery. Nobody had even suspected that the ratio between the lengths of light and darkness could be an environmental factor governing the seasonal behavior of plants and animals. However, by keeping track of the daily changes in the ratio of daylight hours to nighttime hours, seasonal changes could be anticipated.

* The existence of internal clocks seems to have been discovered by the French astronomer De Marian, as early as 1729. De Marian noted that in some plants the movement of leaves which turned toward the sun in daytime continued when the plant was in a dark room. His discovery did not attract much attention at the time.

In their original experiments, Gardner and Allard were working with an artificially bred tobacco plant which had attractive large leaves but, because it would not flower in the field, could not be used for breeding. They found, however, that the same plant would flower in the winter in a greenhouse. Further experiments established that it was the length of the day that determined the presence or absence of flowering. Thus, Gardner and Allard could prevent their tobacco plant from blooming in the winter in the greenhouse by using electric light to artificially extend the daylight period, and, conversely, they could induce flowering in the same plant in summer in the field by shielding it from sunlight for a part of each day.

Photoperiodism turned out to be quite pervasive in life. It is present not only in plant activities such as flowering or seed germination, but also in animal behavior.* Mating, pelt growth, nesting, and insect hibernation are all believed to be manifestations of photoperiodism. The biological clocks which we associate with photoperiodism are influenced by the "light cues" of the environment. Their working can be influenced at any time by changing the relative lengths of light or darkness.

It seems that the mechanisms of the photoperiodic clocks are themselves at least partly governed by circadian clocks, which may be tuned to fixed bright and dark periods. The photoperiodic clocks would then be activated whenever the environmental bright-dark pattern deviates from the fixed one.

It is obvious that both types of biological clocks aided in the process of adaptation by keeping the life processes of organisms in tune with the changes in the environment. These clocks are as accurate as adaptation requires them to be, which is not terribly accurate. They are merely "approximately" or "circa" accurate as their names indicate. But they make it possible for various life-forms to anticipate and prepare for important environmental events, like the approaching of night or the coming of spring. These genetic programs are the first manifestations of foresight and planning in life. And what made this foresight possible was

* Now we think that this process is active in humans as well. It has been suggested, for example, that the relatively new tendency to an earlier age of puberty in developed countries is due to the increased amount of illumination in life.

that these environmental events have always been so regular and so predictable that evolution could read their patterns and program the appropriate responses into the genetic apparatus of even the simplest living organism.

The sky also presents spatial patterns which have either remained the same or changed very slowly during the eons of biological evolution. The apparent path of the sun moving leisurely across the sky is as predictable as are the even more slowly changing patterns of the constellations. Some creatures utilized these patterns by evolving not just biological clocks attuned to periodic changes, but also more complex systems which make very sophisticated use of both duration and periodicity to aid them in spatial orientation and temporal organization. These clocks are often called "continuously consulted clocks" and their manifestations are observable in animal migrations.*

If a migrating bird's behavior were regulated only by the type of clock described previously, it would know when to migrate. But these birds not only know when to go, they also must know where. They must possess internal clocks which help them orient themselves both in time and space. The actual working of these clocks was first observed, not in birds but in ants, back in 1911 by the Swiss naturalist F. Santschi. Later, the orientation of bees was observed in great detail by the Austrian zoologist Karl von Frisch and other researchers. What the orientational abilities of these insects have in common with those of other species is that bees use the sun as a landmark to find sources of food or, on their return, their home base. This is quite extraordinary since the sun is not a fixed landmark at all but something that moves continuously across the sky in daytime. But the bees are not only able to remember the position of the sun but can also correct for its motion. If a bee finds food sometime in the morning by flying at an angle of 30° west of the sun, then around 4 P.M. (say) in the afternoon, it must fly 40° east of the sun to find the same source. Not only can bees do this under normal circumstances, but they can also do it

* These, of course, had been observed and admired from time immemorial. The oldest surviving comment, it seems, is in Jeremiah: ". . . the stork in heaven knoweth her appointed times; and the turtle [turtledove], and the swallow, and the crane observe the Time of their coming home. . . ."

if, between the morning and afternoon flights, they are kept in a dark room. The tiny navigators obviously have a biological clock which not only is a sophisticated chronometer but can also keep track of the motion of the sun. Not bad at all for animals whose brains have no more than a hundred thousand brain cells—a small number compared to our ten billion.

Migrating birds orient themselves in a similar way. G. Kramer of the Max Planck Institute in Germany noted that when migration time arrived, starlings tended to take off at a certain angle with respect to the position of the sun. If the apparent position of the sun was changed by mirrors, the birds tended to take off in a direction which had the same angle relative to the sun's position in the mirror as their former selected direction had relative to the true position of the sun. The birds were also able to orient themselves this way at different times of the day. The experiment was then extended to show that other visual landmarks had no influence on the birds' orientation nor had the switching on of a magnetic field. The latter excluded the possibility that the birds orient themselves to the magnetic field of the earth.

Night migrants seem to orient themselves by using the celestial cues of constellations. It is still an open question, however, whether the night sky offers enough cues for migration or whether other, yet unknown, factors are also involved. In all cases, most orientational abilities are innate—many young birds who have never migrated before are also able to orient themselves with sufficient accuracy.

The champions of navigation are the oceangoing homing birds. In a spectacular experiment by the British ornithologist G. T. V. Matthews, an oceanic bird, a shearwater, was taken from its breeding ground on the island of Skockholm off Pembrokeshire in Great Britain. It was marked, taken by airplane to Boston, Massachusetts, and there released. The bird returned to its Skockholm burrows, about 3,050 miles from Boston, in thirteen days. The shearwater could not possibly have been familiar with the east coast of North America or with any of the features of the western Atlantic. All these are far outside the normal shearwater flying range. Yet it found its way home without hesitation in a very short time. In ironic contrast, the letter announcing the bird's release from Boston arrived ten hours after the bird itself. The shearwater is not unique in its navigating ability. Another oceangoing bird, a Leach's petrel, flew home from Sussex, England to Maine in fourteen days, while a Laysan albatross covered the Pacific from Washington state to its home base on Midway Island, a distance of 3,200 miles, in ten days.*

However complex internal clocks may appear to us, adaptation to a regular and predictable space-time pattern such as that of the cosmic environment is nevertheless a relatively easy task. Living organisms face far more difficult problems in adapting to the small-scale spatial environment. Our immediate spatial environment shows far less order and predictability than the large-scale dynamics of the solar system. The small-scale spatial environment is almost always ad hoc, irregular, and often changes without notice. The problems faced by a living organism in adapting to it are, therefore, much more complex.

Living creatures adapt to space either passively or actively. The passive ones, like most plants, wait for their food to come to them and move only their seeds and spores through space. The active types, like most animals, go out and try to find what they need for survival. Animals are entrepreneurs, and they are all predators who feed on other living things. To find food, animals have to move about actively, observe the external world, and collect information from it. The beginnings of the perception of and adaptation to space and spatial properties must therefore be found in the sensing and processing of incoming information.

These activities probably commenced very early in life's history with one-celled creatures without nerves or any specific sensory equipment. The chemotaxis of bacteria and the behavior of amoebae are good examples. Amoebae, tiny one-celled creatures, move about and help themselves to food by projecting parts of their soft bodies outward. They create, as it were, their own temporary hands and feet, called "pseudopods," which is Greek for "false feet." If, in a laboratory, an experimenter gently touches

* Many nonflying animals can also find their way in large-scale space. Among them are many kinds of fish, whales, sea turtles, and land animals like the caribou. Many of these creatures orient themselves quite accurately while covering enormous distances. Some of these navigational feats also depend on more complex biological clocks; others are accomplished by other means. The homing of the salmon, for example, seems to be directed by the sensitive taste perceptions of the fish.

the side of an amoeba, the creature responds by projecting a pseudopod in the direction of the stimulus. If the touch becomes harsh or too insistent, the pseudopods are withdrawn and the amoeba may also extend a new pseudopod from another part of its body in a direction away from the menacing object. In other experiments, we observe that the tiny creatures tend to move toward regions where the concentration of food is higher and that they try to move away from chemicals which are threatening to their life processes. Therefore senses such as touch or taste, which can give information about the immediate vicinity only, can trigger behavior which already discriminates among spatial directions. In addition, the amoebas can do all this in every direction in three-dimensional space.

But a large-scale exploration of space could only have started with the evolution of the long-range senses—those which can obtain information from more distant surroundings. For most mammals, vision, hearing, and smell are the dominant long-range senses.* In order to be useful, the information obtained by these senses has to be "processed." Irrelevant data must be suppressed, and relevant data interpreted or "understood." This process is exceedingly complex and it cannot be performed without a sophisticated nervous system to act as an information processor. We cannot therefore speak about vision, for example, as a space-exploring sense, before the evolution of multicellular organisms about seven hundred million years ago. Vision, which is to us perhaps our most important sense, is a relatively late arrival in evolution, and more than four-fifths of the history of life had gone by before vision appeared. All was not darkness, however, since long before vision itself, many living creatures had evolved some sensitivity to light. This sensitivity offered some competitive advantages to these creatures by, among other things, enabling them to respond to a light-shadow pattern created by an approaching predator, but it was a far cry from vision.

The development of the ability to see, to collect and process detailed information from afar, was one of evolution's most astonishing achievements. Being able to "see" actually means that an organism is capable of interpreting information contained in the

* There are creatures who evolved senses for static magnetic or electric fields as well. These senses are also long-range but seem to be relatively rare.

incoming light. This is no easy business because in vision the information about the external world is received in an abstract code only. The information may refer to an important object, but it is neither physically nor biologically the same as the object itself. The light reflected from a zebra that reaches the eye of the lion is just a light pattern, i.e., a bunch of photons distributed in a certain manner; it is merely an image of the real thing. And, as R. L. Gregory remarked to make just this point, "One cannot eat an image or be eaten by one. . . ." But, while the light which reaches the lion's eye is neither the zebra itself nor any part of it, it contains information about the zebra encoded in photon patterns. In other words, it is an abstract representation of the zebra. To make use of the information in this code, the lion's nervous system must contain some prescriptions, some programs or processes, which translate the code into a perception of a certain combination of shapes, textures, and colors that together mean "food" for the lion and trigger the appropriate motor response. These programs seem to be mostly genetic but can be perfected by learning.

To perceive an object is then already a very complex process. But the recognition of the zebra is by no means all that has to be deciphered from the optical input. The lion must also know, for example, how far away the zebra is and how fast it is moving. The only information the lion has about distance is the apparent size of the zebra on the retina and the rate at which the apparent size is changing through the relative motion of the two animals. This information is enough to allow the nervous system of the lion to do a fast calculation and decide what to do.

We seldom think of vision as a complex process because, as long as our visual apparatus is unimpaired, we do not have to make any conscious effort to see. But, in fact, nothing is simple about vision. It is perhaps the most complex process in the entire known universe.* Hearing is probably on a similar level of com-

* The enormous complexity of vision has not always been understood. A—perhaps apocryphal—story claims that in the early sixties one of the pioneers of artificial intelligence research gave a student a simple summer project: Solve the problem of vision by designing a model of a computer capable of seeing. After twenty-odd years of intensive research (heavily supported by industry and by the military all over the world), a recent survey quotes an estimate that another ten years could perhaps produce a robot capable of moving over an unknown terrain without stopping for lengthy calculations. And this ability would still be very

plexity. In this case, it is the patterns in the variations of air (or water) pressure that the nervous system has to decode. These are again abstract patterns, they are again "neither edible nor dangerous," and again no part of the source reaches the sensory receptors. Like vision, hearing is an abstract sense.

Information obtained by olfaction seems to be less abstract because, in its primitive form, olfaction is not much more than a kind of highly evolved "long-range taste." The stimuli are airborne or waterborne small parts of the objects *themselves*. Consequently, they refer directly to the biological nature of the object. The messages are not in a code and are neither as abstract nor as ambiguous as in vision or hearing. But the olfactory system became quite important in other respects during evolution. The direction and speed of the motion of the source could be detected with its help, and for some animals, olfaction came to surpass all the other senses in utility. For humans, however, its significance as a space-exploring sense is negligible. Olfaction, like taste, involves sensations which may be pleasurable or not. It may also warn of a danger, like smoke or poison, and it can evoke memories. But the messages remain direct and the brain does not do much work to interpret them. This is perhaps one reason why we cannot create or interpret symbolic forms through either olfaction or taste as we can through vision or hearing and why we have no higher mental activities associated with either of these senses.

With the evolution of the long-range senses, the stream of information from the environment increased tremendously, and

---

far from, for example, the ability to read unknown handwriting. It is quite possible then that even simple acts of visual perception may involve far more complex processes than some very involved logical reasoning. Computers capable of playing competent chess or finding proofs of logical theorems, for example, have already been around for some time.

Current opinion seems to appreciate the sophistication of the visual process and the fact that recognizing a person in a drawing may be a far more sophisticated mental process than understanding a sentence. "The inferential processes that underlie everyday acts of seeing may be deeper and more subtle than are those exercised by an Einstein in the prosecution of his scientific work, but because we have all evolved to be brilliant at seeing but not alas in physics, we tend to marvel only at the latter capacity," write N. S. Sutherland and H. C. Longuet-Higgins in an introduction to the records of a 1979 Royal Society (London) discussion on vision.

its quality was also drastically changed. It became necessary to develop mechanisms able to cope with all the new data, able to process incoming information and make biological sense of it. Thus the most sophisticated structure ever to evolve, the brain, started to develop. The evolution of the brain was parallel to that of the long-range senses, for the one is not much use without the other. Biological evolution, however, seems capable of fostering such parallel developments. The evolution of the sexes is an obvious example. In that case also, one was not much use without the other, yet both evolved.

The power of the brain, human or animal, lies in its ability to develop abstract models of the external environment. The key word here is "abstract" because the faculty of abstraction is what seems to distinguish the brain from all other organs. "Modeling" is a fundamental property of life with or without brains. The very word "adaptation" implies modeling, taking on or complementing some of the features of whatever one is adapting to. The biological clocks, for example, are of help only if their workings are modeled after the environment. An internal clock which triggered behavioral changes every three hours and twenty minutes would not be helpful in adapting to a day-night rhythm. Obviously, all the periodic biological clocks must "model" some periodically changing feature of the internal or external environment. Only then can they help to maintain life's processes.

Some models in an organism may be direct, like the circadian clocks' direct modeling of the Earth's rotation, but the models in the brain are almost certainly indirect and abstract. "Recognizing" a circle, for example, means that we have an internal model, in this case a memory, for the shape of a circle. But it is unlikely that this model consists of a circle-shaped organization of a number of brain cells. It is also extremely unlikely that when we perceive something green, a number of our brain cells actually turn green. The brain does not model the features of the environment directly. Instead, the models are in the form of abstract codes. How such codes are established is not well known even in the case of simple brains. Some codes seem to consist of patterns in brain-cell connections; others are apparently built into certain individual brain cells.

Some of these codes are, furthermore, inborn or genetically

determined; others are established by learning. When a day-old goat or a six-month-old baby tries to avoid, in a laboratory, what looks like a deep precipice, this behavior is obviously inborn. When, on the other hand, a rat, after trial and error, finds its way to the food pellet in an artificial maze, it has learned the spatial model of the maze. It also seems certain that the more developed an organism, the larger the percentage of its "knowledge" that is learned. But, whether inborn or learned, information is obviously stored in the nervous system, and there is some form of internal knowledge of the environment. It is this "knowledge" that we loosely call a "model."

The models in the brain, like all other models in an organism, are useful only if they correspond to the world in which the animal lives. When that correspondence is disrupted, the animal is in trouble. In the late 1940s, Roger W. Sperry of the California Institute of Technology performed a series of experiments on the visual system of the frog. Sperry's experiment made use of the fact that the neurons (elementary cells of the nerves) in the frog's nervous system are able to regenerate. Therefore it was possible to cut the optic nerve of the frog, turn its eyeball upside down, and then reconnect it. The vision of the frog indeed regenerated, but for the rest of its life, the frog struck out for food in the wrong direction. If a fly was moving near the ceiling, the frog responded by striking towards the floor and vice-versa. The frog was never able to modify its behavior. The hapless animal had all the visual information it needed, but, because of its crossed wiring, its internal model did not fit the real world. In addition, the frog lacked the intelligence to adapt to these changes—i.e., to acquire new models which would trigger new behavior.

Humans are much smarter and can adapt to an upside-down world relatively easily. Numerous experiments have shown that if a normal person wears goggles which invert the optical image of the environment (i.e., show the world upside down), after a few difficult days, the person can adapt perfectly to his or her upside-down world and can live a normal life. In the course of such experiments at the University of Innsbruck in Austria, for example, a volunteer student who constantly wore inverting prisms went on a bicycle trip on the fourth day of the experiment and went skiing in the Alps on the sixth day. When the goggles were removed, the subject again experienced a period of difficulty while readapting to the normal perception of space, but soon regained his accustomed orientation.

The animal models of the world developed in tandem with the evolution of the brain. At each stage of evolution, living creatures develop their own models of the external world. These models are, to use Z. J. Young's metaphor, their "cosmologies," and the various degrees of sophistication of the cosmologies correspond to the various stages of evolution; their development parallels the development of the information-collecting mechanisms of the senses and the information-processing mechanisms of the central nervous system.

It was thought for a long time that we would never be able to learn anything about animal cosmologies, about how an animal perceives the world, because we can only perceive our own. But this assumption turned out to be too pessimistic. One of the fascinating results of research in neuropsychology has been the discovery that we could, even if in a limited way, learn something about the "worldview" of some animals. The first successful attempt in this direction was made in the late 1950s by an imaginative team of psychologists at the Massachusetts Institute of Technology: J. Y. Lettvin, H. R. Matturana, W. S. McCulloch, and W. J. Pitts. The cosmology they set out to explore was, again, that of the frog. They found it to be a fascinating world with peculiar spatiotemporal features.

The way to get into the brain of the frog is to make use of the following facts. All sensory organs have "receptor" cells which are sensitive to a certain kind of stimuli. Upon receiving a stimulus, which is always some kind of absorbed energy, the receptor cells send electric signals through the nerve fibers to the brain. In the case of hearing, for example, it is the pressure waves of the air which hit the sensory cells of the ear. They generate electric signals in patterns which characterize the incoming information; the signals then propagate through the nerve to the brain. This process is known, by the way, as "neural coding." The brain cells receive the electric signals and somehow decode them. The details of the decoding processes are not well known even in simple animals, but what is known is that the brain cells involved in the decoding process become electrically active. And this is the important point because that electrical activity, or more accurately speaking, discharge, is measurable. During the 1950s, tiny electrodes, called microelectrodes, were developed, and with their development the electric discharge of a single brain cell became actually measurable in the laboratory.

The classic experiment at M.I.T. was performed using just this technique. The results were published under the whimsical yet significant title, "What the Frog's Eye Tells the Frog's Brain." Lettvin and his colleagues inserted a tiny electrode into the optic nerve of the frog to record its reaction to external stimuli. The eyes of the frog were then exposed to various optical stimuli and the electric impulses from the fibers in the frog's optic nerve were recorded. The results were a revelation. It turned out that the actual "seeing" of the frog was not at all a mechanical, cameralike process. The different groups of fibers did not respond to all light or even to all patterned light. The retina of the frog is much more than the biological equivalent of the photographic emulsion. It processes the incoming information in a highly sophisticated way, amplifying certain features, suppressing others, and separating relevant information from irrelevant. Consequently, there was a response to significant patterns only. Some cells responded, for example, if a small object passed through the visual field. The research team wrote, "We have been tempted to call them "bug detectors." Naturally, the frog's eye needs to respond to small flying objects—bugs. Bugs are a frog's food. Another class of fibers responded when a sudden large shadow fell on their visual field. These fibers might have been called "stork detectors." Detecting suddenly approaching predator birds is just as important for the frog as detecting bugs. Significantly, if the small objects or the large shadows were not moving, the optical cells stopped responding; flies and mosquitoes do not stand still in the air, nor do diving storks.

These examples are enough to make the point. Even a primitive brain such as the frog's can interpret specific stimuli with a well-defined space-time pattern. But the frog's universe like all animal cosmology is determined by evolutionary history. The frog perceives mainly those things which are important for its survival, for which it has developed a model, for which evolution wrote a program. When the frog responds to a flying bug, the process is very different from a mechanical, cameralike response to an optical stimulus. For a camera, the source of the optical stimulus is irrelevant. But the frog could starve while surrounded by an army of dead flies. And it is not because there is anything wrong with its eyes; it is just that its brain is simply not equipped to interpret this particular pattern. The optical pattern of stationary flies presents an alien language to the frog's brain because the frog's brain never developed a model for dead flies. The moment a fly stops moving, it disappears from the frog's universe. For a frog, any fly which stops moving has disappeared into its equivalent of modern astronomy's black hole. And the frog is not even surprised; it does not expect the objects to persist in time and space. The frog has a stroboscopic cosmology; its universe is one of fast-moving light-dark patterns and of stark contrasts: the fly and the stork. The frog, therefore, would not appreciate Impressionist paintings. If it could talk, it would perhaps call them a "senseless confusion, completely mad, grotesque, . . ." using the same words some human critics used on their encounter with Impressionist paintings in 1874. Like frogs, these critics evidently had no models for the Impressionist way of seeing the world. But humans can quickly learn to construct new models. The frog, as we recall from the Sperry experiment, cannot.

3

Adaptation to Chaos

(Space and Time in the Mammalian Brain)

From the amoeba to Einstein is just one step.

Karl Popper

Biological evolution traveled the road from the bland space and time of the protozoan to the considerably more lively one of the frog in about three and a quarter billion years, give or take a few million. The evolutionary road from the world of the frog to the space and time of Newton and Einstein and Rembrandt and Beethoven took, in contrast, only a quarter of a billion years to travel. This acceleration was due to two factors: the development of the mammalian brain in general and the human brain in particular on the one hand, and human cultural evolution on the other. Our concerns here are the evolutionary events that contributed to the development of the peculiarly mammalian and human capacity for always perceiving the world in space and time. A few years ago, Harry Jerison of the University of California at Los Angeles put forward a theory that gives a possible explanation of how and why these events came to pass.

The working assumption underlying this theory is that the main function of the brain throughout evolution has been to make biological sense of the stimuli it receives through the senses. The stimuli come into the brain in the form of nerve impulses. We say that the brain makes biological sense of this information, or processes this information in a biologically useful way, if in these stimuli it detects information that is useful in adaptation. An efficient way to do this is to construct an internal model of some relevant features of the external world. A frog's brain, for example (as in the case of the M.I.T. frog), processes information in a specifically "froglike" way, and builds itself a model of a "stroboscopic" universe. This model enables it to catch flies or to escape storks; in other words, it makes biological sense of the information the frog receives, and thus increases the frog's chances for survival. The question now arises of how these mechanisms worked in later evolutionary history and, in particular, in human evolution.

It has long been known that the general trend in the evolution of the brain has been a gradual increase in size. Size alone, however, is not a very useful measure of the overall development of the brain. Larger animals obviously need larger brains to maintain order in their bodies. But this does not mean that they are necessarily more intelligent than smaller animals. A more relevant measure for characterizing the degree of development of the brain of mammals seems to be a number called the "encephalization quotient." For any species, this number is the ratio of the brain size of that species to the average brain size of mammals having the same body size. Therefore, when we speak about the enlargement of the brain, we always mean the increase of the encephalization quotient.

Jerison suggests that the enlargement of the brain went hand in hand with the evolution of an internal model of the external world in which permanent and separate objects were perceived in space and time. Obviously a large and complex brain is necessary to perceive the external world as consisting of "objects in time and space." But beyond that, Jerison argues that it was the creation of an internal model of the world as objects in time and space that in itself promoted the evolution of the brain and of intelligence.

Jerison follows the evolution of the brain from the beginnings of its recorded history about, as he puts it, "425 million years ago when jawless armored fish living in fresh water estuaries left their heads to posterity." Jerison bases his theory on paleoneurological data (paleoneurology is the branch of paleontology which extracts information about the nervous systems of extinct animals from their fossil remains). For our purposes, it will be sufficient to start in an era when something like the frog's view of the universe was the most advanced cosmology—the era of amphibians and reptiles. More accurately, we begin the history of the evolution of the brain at a time when the invasion of dry land by reptile-like vertebrates took place. This event—the beginning of vertebrate land life—took place about a quarter of a billion years ago in the geological period we call the Mesozoic.

The reptiles adapted to land life in various ways. Their main long-range sense was vision. Early reptilian vision, as we assume from observations of present day reptiles, was very different from the later, more complex mammalian visual sense of which human vision is one example. The visual information in the reptilian eye was processed mainly in its retina, which was rich in light-sensitive receptor cells, and the small brain of the animal did not play any major role in interpreting or processing visual information. Reptilian vision was, on the whole, probably similar to the frog's. It could not follow the motion of objects for any length of time, nor did it work well at twilight or in the dark. But it was quite adequate for the reptiles, who were, in any case, inactive during the night.

The earliest mammals were small animals who evolved as an offshoot of the reptilian adaptation to land life and who were, at the beginning, not very different from their reptilian predecessors. These new creatures were forced to become active at night in order to hide from their dangerous cousins, the large reptiles who ruled the dry land during the day. The "cold-blooded" reptiles could only be active when the sun warmed their bodies. In contrast, these early mammals already possessed a heat control mechanism which allowed them to be active at night. But when these mammals began to evolve into nocturnal creatures, their vision was still of the reptilian, daylight variety, and of course, was not of much help to them in the darkness of the night. As a result, they could only survive if they replaced vision with hearing and/or olfaction as their main long-range sense.

Being forced to use hearing or olfaction instead of vision to obtain information from a distance had a twofold effect on the mammals' brain. The first, Jerison argues, was that a whole new system of interconnected neurons along with a large number of brain cells capable of interpreting auditory and olfactory information had to be "packaged" into the brain. There was simply no room elsewhere on the periphery. An elaborate visual apparatus could evolve around the retina, but in the auditory and olfactory systems, there were no analogous places large enough to accommodate the neural apparatus necessary to interpret the information coming from these new long-range senses. Thus a new processing system, a new neural circuitry, had to evolve within the brain, making the brain larger. The second effect, although once again involving an enlargement of the brain, involved, in addition, the evolution of a more sophisticated brain. This latter effect has to do with the nature of auditory (and olfactory) information.

For most of us, vision is the main distance sense, and we therefore have to stretch our imaginations to appreciate the ways purely auditory distance information can be interpreted. Olfaction is only a weak, rudimentary sense in humans (few of us, for example, could perceive through sense of smell the motion of a distant object), and for this reason we will consider it in passing only.

Hearing works differently from vision as a distance sense. Vision is well suited to handling information coming simultaneously from various sources in different places. Hearing, in contrast, is better at handling information extending not in space but in time. Auditory information comes in temporal successions. This is fine when we listen to music or speech or, generally, when we need information about temporal order. But temporal information is not what a nocturnal animal needs. What it needs above all is spatial information. Temporal information is only useful to the extent to which it contributes to the animal's knowledge of the spatial situation or to a construction of a spatial "map." This cannot be done unless the animal is *capable of translating a temporal sequence into a spatially organized order or map.* Such a map depicts the spatial behavior of the source of the stimulus and is able to answer biologically relevant questions such as Where is it? Is it moving? In what direction? How fast? To get all this information from a sequence of sounds is difficult. Jerison uses the following example to illustrate the problem.

> If we try to picture a vista from a set of brief exposures to successive fractions of the total scene, with a second or more between exposures, we see the problem of using temporal information when spatial integration is the more "natural approach." To put it more simply: in order to interpret auditory information from a distance, the brain has to develop the ability to "translate" the temporally encoded patterns of auditory nerve impulses into the equivalent of spatial maps.

This is no easy matter. A mechanism able to do this must be as sophisticated as the TV camera which performs approximately the same task. Biological mechanisms of such complexity do not develop easily. But the evolutionary advantage went to those creatures whose brains became large enough and complex enough to be able to utilize auditory and olfactory signals to correctly perceive, for example, a moving predator. This kind of information processing involves the use of time intervals to represent spatial dimensions. It is reasonable to assume that this was one of the more significant ways in which "time" as a dimension of existence first entered into the interpretive processes of the brain.

Around the same time as their auditory and olfactory senses were developing, the reptilian vision of the early mammals was also being pressured to change and evolve. Gradually, such features of later mammalian vision as the ability to see in twilight or moonlight began to develop. Mammals were thus receiving information from the same source through two or sometimes three different sensory modes: vision, hearing, and olfaction. It was important to learn how to reconcile all these perceptions and this also held new consequences for the brain.

In order to make sense of the complex array of data, it became necessary to perceive the source of the different sensory experiences as being "the same." Therefore, the evolution in the brain of some specific neural connections, or some kind of code, was needed; the code would integrate the information received in different sensory modalities and *interpret it as coming from the same spatial source*. It was this kind of sophisticated processing that opened the way to a new perceptual world. What were before merely incoherent patterns of unrelated stimuli now became permanent sources of information, or "objects." Thus the perception of permanently existing objects in space and time originated. The evolutionary advantages of such a development will be discussed below. But it is clear that from this point on, mammals lived in a perceptual world which was becoming increasingly different from that of the frog. The frog's brain, as Jerison points out, does not use a model of permanent objects in space and time as an intermediate stage between stimulus and response. The frog's brain reacts immediately to any stimulus. If the same fly happens to pass several times through its visual field, the frog merely sees this as a series of disconnected events, and it acts on each event independently. Its brain is a "complex reflex machine."

A second and entirely independent acceleration of the evolution of vision occurred, it is believed, at the end of the "era of the reptiles" about sixty million years ago, when the dinosaurs and the other ruling reptiles, together with a great many other marine and land organisms, became extinct. The causes of this sudden and dramatic mass extinction are not known with certainty. It is believed by some that it was caused by the effects of some extraterrestrial factors on the environment, the fall of a huge asteroid perhaps, which changed living conditions so abruptly that many life-forms were unable to adapt and perished. Jerison demonstrates at this point, incidentally, that contrary to scientific folklore, the dinosaurs were not small brained but had a normal-size brain by reptilian standards. Whatever happened, we are among the long-term beneficiaries of this cosmic accident, because the extinction of the ruling reptiles opened up new opportunities for mammalian life. After the disappearance of the big reptiles, the lowly mammals suddenly found themselves in a more friendly environment, with less competition and danger in daytime, and many of them abandoned the nocturnal lifestyle and became active during the day. And as a result, it again became advantageous to see well in daylight.

The adaptation of the visual system to this new demand was not, however, a mere retrogression to the old daytime reptilian vision. By that time, mammals already possessed fairly well-developed brains with advanced mechanisms for the processing of auditory information. Jerison argues that it was simpler and more efficient to model the new visual system on the already developed auditory one than to restore the—by then lost—mainly retina-based reptilian vision. This assumption is plausible on general evolutionary grounds. When the need arises for new functions, evolution seldom creates new structures when it can modify existing ones. The latter is achieved much more easily because simpler changes are more likely to happen than complex ones. François Jacob of the College de France explained this: "Evolution," he writes, "resembles not engineering but 'tinkering.'" An engineer "works according to a preconceived plan. . . . He has at his disposal materials and machines specially prepared for that task," and "the objects produced approach the level of the technology of the time." Biological evolution, in contrast, "does not produce novelties from scratch." It has no well-defined long-term aims, has no specially prepared materials; it works with whatever already exists, and it uses whatever odds and ends are available. The evolutionary process behaves "like a tinkerer who, during eons upon eons, would slowly modify his work, unceasingly retouching it, cutting here, lengthening there, seizing the opportunities to adapt it progressively to its new use." In such a process it was more likely that the new mammalian vision evolved to be not something radically new, but rather a mechanism modeled after the already well-developed auditory and olfaction systems.

Because of this, the "new" vision could not only do space mapping but, like audition, could handle extended temporal sequences as well. It could integrate those stimuli which arrived over longer time intervals into the perception of a single event. This ability is called "timebinding." An example of timebinding in audition is the human perception of a succession of disjointed sounds as a melody. In visual timebinding, an animal is able to follow something moving across its field of vision and perceive it as a single object. In order for the animal to do this, the brain has to be able to store visual images, to retrieve and compare them with later images, and finally to evolve a code which "labels" the sources of all these images from different times and different points in space as being "the same." All these complex functions require new neurons, more neural connections, more brain tissue—in short, a larger brain.

As we shall see, there were major adaptive advantages in all this complexity. The mammalian brain received, through its diverse senses, a much greater amount of information than did the lower animals. As a result, the "world" of the higher mammals became not only much richer but also very irregular and rapidly-changing. It was much more difficult to adapt to such a world than to one with only such regular or slightly-changing features as day-night rhythms or normal variations in temperature. Successful primitive organisms had already adapted to these, but it was only the more developed animals who could adapt to a world of constant and irregular change. But this shower of information impinging on a mammal's senses is, taken by itself, almost completely disorganized and devoid of obvious content. To understand this, we merely have to imagine what would happen if we replaced our sense organs with an array of electronic devices which counted incoming photons, registered variations in air pressure, measured changes in the chemical composition of the floating molecules in the air, and so on. Let us further imagine that after a short time we wanted to see what information these devices had recorded. We would, of course, find largely chaos on the records. Since the devices simply record indiscriminately everything that hits them, they record the random noise of the external world. But it is precisely this random noise, this cacophony, which reaches our senses and those of any mammal; the random noise is the raw material which is received by our nervous systems in every instant of our lives. The advanced animals are therefore faced with the very difficult problem of making sense of the bewildering noise coming from their surroundings, and of adapting to this apparent chaos.

In order to survive, an animal has to make biological sense of the chaotic stream of stimuli. It has to detect information in the apparent disorder, it has to discover patterns in the noise, and it has to process whatever information can be found in a biologically meaningful way. It seems that, in mammals, adaptive pressures resulted in the evolution of a particular method of information processing. This method is based on the fact that the handling of a continuous flow of information can be simplified by breaking the information down into smaller, more manageable pieces, and organizing them into a coherent system. This is precisely what the mammalian brain does. It breaks down incoming stimuli into aggregates which are then labeled in the brain as "objects" and organizes them into the perceptual mode of space and time. The great adaptive value of this mode is that it gives structure to the perceptual world, making it stable and reliable, and thus making adaptation possible.*

This framework, the cosmology of "permanent objects in space and time," enables the brain to organize information with the help of such processing "subroutines" (to use computer language) as the so-called *perceptual constancies.* An example of such processing is what happens when we see a person walking away from us. The person casts a series of smaller and smaller images on our retina, yet we see neither successive persons of different sizes nor a single person shrinking. To take another example, when we turn our head and the whole visible world moves across our retina, we do not feel the outside world moving. There exist powerful, mainly innate mechanisms in the brain which correct our perceptions and allow us to perceive the size of a receding object as constant and the outside world as being at rest.

There are obviously many such built-in mechanisms in our brain, and they are of extraordinary importance. These are the mechanisms—prewired evolutionary programs as simple as they are ingenious—which allow us to keep the outer world stable and reliable. They enable us to interpret a change in "sensory input" from the external world (for example, a changing image on the retina or changes in what we hear caused by the movement of a sound source) as *a change in our perception and not as a changing external world.* We seldom consciously notice this particular faculty of our brain because these mechanisms are built into our nervous systems and they work automatically. But this extraordinary capacity of the brain makes the chaotic external world stable and predictable. It seems then that the mammalian brain organizes its sensory input, its perception of the world, in the framework of space and time in order to be able to make sense of it. The framework of space and time simplifies the world, gives it a coherent order, and thus makes it liveable.

* The way we use hearing offers a simple example of "object forming from chaos." We stand in a room filled with the sounds of people conversing. A recording, made with a sensitive microphone which registers the same slight changes in air pressure our ears do, plays back a perfect cacophony. Yet when we are there, we can converse with a neighbor because our nervous system processes the auditory input in such a way that we can select only those sounds which are of interest and disregard the rest. Vision can do the same type of processing when the need arises.

The inborn models of the mammalian cosmology are not necessarily ready for use at birth. These models have to first be trained, activated, and stabilized. They are often potentialities more than ready-made plans for action. It seems that even the very perception of space and of patterns and objects in it has to be learned before the inborn models can function properly. A German physician, M. von Senden, collected some case histories of patients who were born blind but whose vision was later restored with cataract surgery. It turned out that these patients could not, at first, interpret their new visual sensations at all. They were not able to judge distances; they could not recognize shapes and sometimes failed to recognize the distinctiveness of objects. Nor could they correlate their visual and tactile experiences. Most younger patients learned all this after some time, but most of the older ones had great difficulties in coping with their visual world even after a long time.

There is also evidence that in newborn animals the perception of simple spatial patterns needs training before it becomes automatic. Colin Blakemore of Cambridge University kept newborn kittens in a visual environment where everything was striped, either vertically or horizontally. When they grew up, these animals were quite able to recognize vertical or horizontal stripes but were very poor at recognizing any other visual patterns. The human infant, to take another example, also has to learn a great deal before it knows how to use the inborn models of its brain. According to the experiments of the late Jean Piaget of the University of Geneva, for example, an infant does not know about such a fundamental feature of mammalian cosmology as the permanence of objects until he or she learns it from experience. The ability to perceive permanent objects is usually not established before the child is close to a year old.* The message here is that the human perceptual cosmology must, in a sense, be learned by each child, but the overall character of this cosmology is predetermined for the species as a whole.

* Piaget also remarked that while each reader's own child always learned everything at a much earlier age, the average was maintained wherever such experiments were performed.

All this has interesting consequences, not only for the interpretation of time and space but also for the continuing debate over the old philosophical riddle: How does the human mind learn about the external world? When discussing this question, most scientists—i.e., most of those who have bothered to think about it at all—accepted, at least until quite recently, the *empiricist* view. The empiricists suggested that everything we know is learned from our individual experiences. The human brain, according to this view, comes into existence as a clean slate. It is blank—contains nothing—and only by accumulating experience does it come to hold information about the world. We learn as we go along. A good many philosophical schools have agreed on this one point, even those which disagreed on most other things. Since the Renaissance, such otherwise diverse philosophers as John Locke, David Hume, French encyclopedists like Denis Diderot and Jean d'Alambert, as well as Karl Marx, Ernst Mach, and Bertrand Russell, have, among others, shared this view. The idea that the human brain could have innate properties—other than its ability to think—or that it might contain inborn models of the environment which have no connection with the experiences of the individual, was held by the empiricists to be nonsensical.

Scientists probably found this view attractive because it proposed a model for the development of the brain which paralleled the development of science and the scientific method. Science itself has, after all, evolved through observations of, and experiments with, the external world. In fact, the increasing primacy of experience over speculation and over prejudices was a fundamental feature of the evolution of science.

Some philosophers, however, opposed the empiricist view from the earliest times on. These thinkers maintained that much of the order we perceive in the external world is, in fact, the product of our own mind. Such *rationalist* views have been held since the Renaissance by René Descartes, Baruch Spinoza, Gottfried Leibniz, and Immanuel Kant, to mention only the most distinguished. Kant in particular proposed that the perceptual framework of space and time itself was a property of our own mental apparatus and that, therefore, it made no sense to speak about space and time as being objective properties of the external world.

The motivations of rationalist philosophies were sometimes religious. Some of the rationalists argued that if the human mind had inborn properties, those properties could have been put there only by supernatural agents. It was these religious associations that were yet another reason why many scientists rejected the rationalist schools of thought. It is therefore ironic that the modern theory of evolution—a theory not usually viewed with favor by most organized religions—does lend some support to the ideas of the rationalist thinkers. One day, perhaps, we will know to what extent their views are correct; we certainly do not know now.

It does seem certain that when the brain creates internal models, these must conform, at least in part, to the real world. We do have inborn models for time, and for space and the objects in it, but these models were formed by the process of adaptation to a real world. A mountain goat is probably born with a ready model of precipices in its brain, but this does not make the precipices of the external world any less real. In fact, a goat which behaved as if the precipices were "only in its head" would have little chance to propagate such beliefs through its offspring. The mammalian brain organizes incoming stimuli according to its internal models. These can be biologically successful only if there are things in the external world which correspond in some way to this particular mode of organization. Therefore, even without knowing much detail about the relations between internal models and external world, it is possible to conclude cautiously that the empiricist and rationalist views can, to some extent, be reconciled. The empiricist doctrine that all knowledge comes from experience is essentially true. But the rationalist view can be accommodated by adding that the word "experience" does not mean only individual experiences. It also includes genetically encoded experiences or the knowledge of the environment that we acquired in adapting to that environment during the course of our long biological past. Kant, in other words, may well have been right. It seems likely that the perceptual framework of space and time is indeed a property of the mental apparatus with which we are born. But this apparatus came into existence during hundreds of millions of years of adaptive learning, and it was this long learning process that determined the basic features of our "real" world—the only world we can ever know.

As a final example, consider a freshwater pond in which amoebae, frogs, and beavers live together. The physical environment is the same for all these creatures, yet the worlds they perceive, their internal models, are as different as their evolutionary pasts. The amoeba probably experiences a rather simple world of essentially two alternating features: the pleasant and the unpleasant. It reacts to them by moving toward or away from the perceived source of the stimuli. The frog's model of the world consists of fast-moving shadows and it reacts to them with a reflexlike snapping and jumping. The nervous system of the beaver experiences the world of the mammalian cosmology—the beaver's model of the world consists of permanent objects in extended space and time, and it organizes its behavior accordingly.*

Mammalian cosmology reached a new stage with the evolution of human beings. For not only do humans perceive objects in space and time, they also create symbols for "objects," for "space," and for "time." With the use of human symbols, the cosmologies of symbolic space and time were created. The most important symbols are, of course, words themselves: the symbols of language. Language allows the human brain not only to perceive objects and events in space and in time, but also to represent them as concepts, think about them, and communicate those thoughts. Perceiving and thinking have at least one important property in common. When the modes of time and space evolved in the mammalian brain, one of their roles was to interpose themselves between stimulus and response, to allow for the evolution of other than reflexlike responses. One of the roles of language is quite

* The idea that although all creatures live in the same external world, they experience only their own special one was first suggested by the German biologist Johann von Uexkull in 1908. For some reason, however, von Uexkull did not see this as a result of adaptation but believed in some special force which created the animals' inner models to correspond to the environment.

A more modern approach, and the one we adopt here following Jerison, was pioneered by the late Kenneth J. W. Craik of Cambridge University. Craik suggested that the brain constructs models of the external world using neural processes that symbolize the workings of the external world in an abstract form. These models "explain," or make sense of, the incoming information and bring about a reaction from the nervous system. Since these models represent the "real" world for a creature, different brains create different worlds.

similar. Like time and space, conceptual thinking also interposes itself between stimulus and responses, creating a still larger variety of responses. This analogy suggests that, in some respects, the evolutionary role of language may have been similar to that of the internal models of space and time. This, in turn, raises the question of whether the origin and evolution of language might be explained in somewhat the same way as we have explained the evolution of time and space in the brain. Jerison thinks that it can.

There are few well-established facts which could be of help with the problem of the origins of language and human symbolism. What seems to be known is that in the evolutionary branch which led from nonhuman primates to present day humans, brain size increased at an unusually fast rate. During the last three million years, brain size on this branch has more than tripled. This growth accelerated particularly strongly during the last half million years, which is a very short time span for such large-scale changes.* In contrast, during the last hundred thousand years, the size of the human brain has not changed at all.

Jerison argues that all these facts together with the evolution of the human capacity for symbolic language and of human culture can be explained in the framework of his theory. He suggests that the same processes through which the mammalian brain evolved continued to operate and thus led to further increases in the size of the brain. In particular, the process of integrating information coming from different senses was further refined, leading to an even more precise perception of objects. To facilitate this, the mammalian auditory system had to evolve further, modeling itself, in turn, after the already well-developed mammalian visual system. This meant in practice that it became possible for the brain to convert not only visual images but also sound patterns into "permanent objects," and this ability made possible the auditory location of objects. In this now even more advanced perceptual world, objects could be perceived sometimes by seeing, sometimes by hearing or by smelling, and still be identified as

* In the evolution of the brain of the ungulates, for example, we find a doubling of size in about ten million years, which means that the process required 2 to 3 million generations. The early humanoids, it seems, doubled their brain size about two million years ago in a hundred thousand generations. This is about a twenty to thirty times faster growth.

being the same. The world, in other words, could now be perceived in a coherent spatiotemporal framework in which objects could be seen, heard, smelled, and touched, and their temporal permanence was ensured. The mental representation of objects could now be stored in the nervous system and objects were now remembered. The integration of different sensory images was no longer momentary but could be extended over long time spans, over days or weeks, perhaps over years.

We can also assume that at some point during this evolution, it became possible to break up the memorized images of the past into parts. These could then be recombined to form new images, wholly new configurations which had never been experienced before. Thus a most extraordinary abstraction evolved. It became possible to imagine events which had never taken place as well as to recall those which had actually been experienced. It was imagination which made *conscious* planning and foresight possible—which precipitated the creation of what became a most important symbolic time: the future. These new capacities, memory and the ability to conceptualize the future, had their own evolutionary advantages, and they also required more brain tissue, a larger brain.

As the evolutionary process continued, the integration of sensory information was extended from objects to images. It became possible to evoke visual images by auditory signals. This was advantageous because in this way the visual images could be labeled using an auditory code, a set of symbols which could be used to reevoke the images and allow them to be recognized at their next appearance. If, perhaps by a lucky coincidence or perhaps by the logic of adaptation, this happened to a species which was endowed with the motor ability to utter a variety of sounds, then this development opened up a whole new kind of auditory coding of objects and events and their images, and, in this way, the stage may have been set for the beginnings of speech, of symbolic language.

If this hypothesis about the origin of language is essentially correct, then the ability to communicate was probably of secondary importance in the very early development of human symbolic language. More important was the ability of symbolic language to aid in cognition and imagination, to produce new models, and thus to create another powerful way of making sense of the outer world. In discussing this particular question, Jerison points out that the ability to communicate is widespread in the animal kingdom. Individuals in many animal species communicate with each other whatever information is necessary for survival and propagation. But this communication mostly takes the form of "fixed action patterns," like the wagging of a dog's tail, in which the meanings and the forms of the signals never change. The signals of animal communications are mostly rigid, and are predetermined, genetically programmed into the animal's nervous system. The ability to communicate, therefore, is not a sign of higher intelligence in itself, nor does it require large amounts of brain tissue. In later human evolution, language was used for communication. This became by far the most important factor in the further development of language ability. But its original function, the chief reason for its emergence, according to Jerison's theory, was to aid in cognition, to evoke recognizable imagery, and to label auditory and visual images. "We need language to tell a story much more than to give direction for action," writes Jerison.*

We can visualize the sensory integration process by invoking the biological principle of recapitulation. This principle was formulated by Ernst Haeckel, a nineteenth-century German biologist, and it says that, during its individual development as an embryo, an animal goes through the same stages its ancestors did during evolution. The individual development recapitulates the evolution of the species. The general validity of this idea has often been questioned, but it is a fact that all human embryos, for example, go through stages where they resemble first fish, then reptiles, then nonprimate mammals before reaching a human shape. In the spirit of this principle, we can imagine that a short replay of sensory integration and sound labeling actually occurs in each human development. This might happen approximately along the following lines. As a very young infant, you saw your mother's face, heard her voice, sensed her smell, touch, and taste.

* Linguists had expressed such ideas earlier. The noted American linguist Edward Sapir wrote in 1932 that ". . . the purely communicative aspect of language has been exaggerated. It is best to admit that language is primarily a vocal actualization of the tendency to see realities symbolically, that it is precisely this quality which renders it a fit instrument for communication and that it is in the actual give and take of social intercourse that it has been complicated and refined into the form in which it is known today."

It is unlikely that you "knew" immediately that all these different sensations came from the same source, but these sensations were gradually integrated by your nervous system. At a certain point, the voice of your mother, for example, evoked the image of her face, her touch, her taste, and so on. In a few months' time, you were able to respond to the auditory label of the common source of those sensations and you probably smiled when you heard the word "mummy." In another few months, you learned how to evoke the sensations yourself by pronouncing the very same word.

The immediate advantage of labeling is that when objects, events, and images are precisely labeled, as long as the labels themselves are remembered, the objects will always be recognized. Labeling, remembering, controlling the motor apparatus, regulating speech, evolving rules of syntax and the rest, required an enormous number of new neural connections and a huge amount of new brain tissue. It is therefore not unreasonable to assume that the explosively fast increase in the size of the human brain during the last half million years resulted from the adaptive advantages of the development of language and of other capacities for imagery and memory.

The larger language-capable brain constructed a human cosmology that was very different from the nonhuman. The appearance of symbolic language was an evolutionary event of the greatest importance. It broke the monopoly of the process of biological evolution in life. Before language appeared, virtually all important information relevant to life which had to be passed on to the next generation was encoded in the genes—in the spatial structures of the DNA molecules. There was no other way. When an animal died, virtually all of its acquired knowledge died with it. The evolution of symbolic language changed this. Information from then on could also be encoded in temporal sound structures like words and sentences.

As a consequence, as far as the human race was concerned, the laws of biological evolution, in force for three billion years, became relatively less and less important in comparison with the few millions of years of cultural evolution. The temporal code of language enabled humans to transmit vast amounts of individually acquired knowledge to the next generation. This meant that to acquire useful new information it was no longer necessary to wait eons and eons for accidental genetic changes. The information could now be obtained from the previous generation without much delay. As a result, cultural evolution progressed at a much faster rate than biological evolution ever could. The human heritage began to consist of two parts: one that had been amassed during the whole evolutionary history of our species and is still encoded in the molecular structure of our genetic material, and another that was accumulated during the last couple of million years and is encoded in our language and in other symbolic structures.*

The rate of progress of cultural evolution became even faster when, with the evolution of drawing and writing, information was again transmitted via a spatial code. This code, however, unlike the genetic code, was stored entirely *outside* the living body. Thus, a new way of storing information had been created which later became the dominant mode in technologically advanced societies. In our time the extrasomatic (outside the living body) storing of information is an increasingly important part of individual and social life. What we store now is not only in writing but in many other information-preserving forms like films, audio and video tapes, musical notes, computer programs, and pictures. As Karl Popper often points out, an increasing part of human knowledge will, in the future, be stored outside of the human brain.

All this is the result of the extraordinary ability of language to build symbolic world models. And these models help make sense of the world in an entirely new way. In a new way but in the same sense as when, for example, the models in the brain enable an organism to *see,* i.e., to make sense of the information encoded in the patterns of incoming light. It is now generally believed, following the suggestions of Noam Chomsky of the Massachusetts Institute of Technology and the late Eric Lenneberg of Harvard University, that, not unlike the ability to see, the ability to use symbolic language is also an inborn capacity of the human brain.†

* Symbolic structures are not the only acquired information that we can pass on to succeeding generations. Man-made tools and the skills involved in their use also constitute information. These are neither genetic nor symbolic. Yet they are important parts of our heritage.

† Like the ability to see, it is also an inborn capacity which has to be "activated" before it is ready for action. No one can use language without first being exposed to one.

Yet, in spite of such similarities, language works differently from perception. What we perceive are always patterns either in space or in time, or in both. Language transcends biological space and time and offers new patterns: *patterns in meaning*. You can classify sentences by deciding which of them are true and which are false. Neither truth nor falsehood is commonly organized in space or time. There are also grammatical patterns of meanings. You can ask, you can deny, you can assert. And there are the concepts which symbolic language has created, such as numbers. Numerical concepts such as "even" or "odd," for example, are neither in space nor in time. These, like many grammatical patterns, are cognitive and can be grasped through conceptual thinking only.

An important aspect of the transcendence of the senses by symbolic language and other symbolic forms is the evolution of "symbolic time and space." This uniquely human creation provides a framework for our mental activities both in science and in the arts. I shall come back to this issue presently, after a short digression into the meaning of symbolism.

The general problem of human symbolism has occupied many twentieth-century thinkers. Among the philosophers who have studied it are Ernst Cassirer, Nelson Goodman, Suzanne Langer, and Alfred North Whitehead. One of the thorniest questions one encounters in these studies—and it is a question which has never been satisfactorily answered—is how to define the exact nature of a "symbol." A useful approach to this problem seems to be to ask about the difference between what is called a *signal* and a real *symbol*.* A signal—tautologies can sometimes be useful—simply signifies something. The smell emanating from a kitchen signifies that something is cooking. The sound of the doorbell signifies that somebody is at the door. The wagging of a dog's tail signifies that the animal is happy. All these are "signals" and their meanings can be as clear to humans as to animals.

Words such as names can also be signals—but they can also mean more than that. "If," writes Suzanne Langer, "you say 'James' to a dog whose master bears that name," the dog will interpret the sound as a signal and will "*look for* James. Say it to a

* Both signals and symbols belong, according to currently accepted usage, to a more general class: *signs*. A sign is anything which contains information.

person who knows someone called thus, and he will ask, 'What about James?'. That simple question is forever beyond the dog . . ." The name, in other words, does not mean more for a dog than its owner's smell or the sound of that person's footsteps. It signifies the person to the dog. "In a human being, however, the name evokes the conception of a certain man so called, and prepares the mind for further conceptions in which the notion of that man figures; therefore, the human being naturally asks, 'What about James?' "

A signal, writes Langer elsewhere, "indicates the existence—past, present or future—of a thing, event or condition." All we can do with a signal is to "react toward it" as the dog which started to look for James did. "Symbols," in contrast, "are not proxy for their objects, *but are vehicles for the conception of objects.*" Or events or conditions, one may add. "To conceive a thing or a situation is not the same thing as to 'react toward it' . . . or to be aware of its presence." In talking *about* things, we have conceptions of them, not the things themselves; and it is the conceptions, not the things, that symbols directly mean.

Animals seem to communicate by signals only.* Humans use signals as well as symbols. When we touch something burning we can exclaim "Ouch!" or we can say "Fire is hot." The difference between the two is essentially the difference between a signal and a symbolic expression. The first, in the words of George Gaylord Simpson, is "an interjection expressing the affect of the moment," the second is "a symbolically coded representational communication of a timeless generalization."

To summarize, signals carry primitive meanings. They always refer to something physically present, perceivable. They never go beyond the level of communication, they cannot be generalized or combined to refer to something else, and they can refer to a limited number of specific situations only. Symbols also carry meanings and can also refer to something perceivable. The same expression (like "James") can be a signal or a symbol, depending

* The controversy surrounding the question of whether chimpanzees can or cannot be taught to use symbolic language is slowly dying down. Many experts seem to have come to the conclusion that these intelligent animals can cleverly manipulate signals and use them to communicate their desires. They are nevertheless very far from mastering even the rudiments of symbolic language.

on circumstances. But symbols can also refer to other symbols. This latter ability makes the range of human symbols not merely large but infinite. There is no limit to how far symbolic forms (e.g., sentences, numbers, shapes, sounds) can be generalized, combined, and extended. Symbols can refer to any aspect of the real world, perceivable or not, as well as to infinitely many abstract or imaginary worlds. They can abstract to things, situations, and events not present; they can refer to the here, the elsewhere, the past, the future, the never has been, and the never will be.

To establish the difference between signals and symbols is useful in appreciating the difference between what we can call "biological" and "symbolic" time and space. Like signals and symbols, the two can overlap. But the biologically relevant time and space is the here and now only. The here and now is all we can perceive or can "react toward." We cannot perceive or do anything in times past or future or in places outside of our sensory experiences. Therefore, the biologically relevant time and space is restricted and small. It can be extended somewhat, in the case of more evolved creatures, to include some memories from the animal's own lifetime and some expectations based on remembered experiences or on genetically encoded memories. But that is all. No dog knows, as Kenneth Boulding puts it, "that there were dogs before him and there will be dogs after him." Nor do the cats in Rome know about the mice in Athens.

The human world is very different. It also includes the here and now, but it includes in addition the infinite worlds of symbolic time and space. It includes "yesterday" as well as the "time when the sun will die." It includes the "center of the most distant quasar" and "the average position of a proton in a hydrogen molecule." It includes the day when the now newborn will first go to school as well as the "first three minutes of the universe." None of these are relevant to the here and now, none of them are perceivable, and all are made up of human symbols like words, concepts, numbers, geometrical figures, and so forth.

The mental worlds of the very far and very small, of the distant past and future, have no limits. Yet they do not even begin to exhaust the varieties of symbolic times and spaces. The space in a painting is neither large nor small. It is always symbolic. You cannot walk into it or experience it by being in it. It is "space" for the human mind only. The same is true of a map of Antarctica, of

Hilbert space (a mathematical construct such as, for example, a space of square integrable real functions), or the space where Pac Man moves, or the Garden of Eden for that matter. All are symbolic spaces of the mind. Symbolic times also come in infinite varieties. We can read Leo Tolstoy's *War and Peace* in days, but our mind experiences all the symbolic years in which the story unfolds. Watching a movie, listening to music, reading about history, solving a time-evolution equation in physics, all involve the experiencing of symbolic times. They are all different from each other and all enlarge the perceivable now.*

As all human beings think in terms of symbolic time and space, it is natural to ask about the evolutionary significance of these mental constructions. To ask, in other words, what adaptive needs could have been served by the appearance of such seemingly irrelevant ideas as the long-vanished past or the distant future? What was the biological usefulness of imagining places one had never seen? We can, of course, only speculate on such problems, but we should do so in the spirit of conservatism. In other words, we should assume that the role played by symbolic time and space in the human symbolic world was similar to the one played by perceptual time and space in the perceptual world. As perceptual space and time was useful in processing "direct" sensory information, symbolic time and space must have been needed to process information in this new mode, information which, while also coming through the senses, was now encoded in symbols.

Perceptual time and space was useful, for example, in interpreting and integrating inconsistent sensory input. Imagine a primitive creature endowed with primitive sensitivity to light-dark patterns. The creature perceives from a signal the "presence" of an object. Attempting in a reflexlike fashion to touch it, however, it finds nothing to touch. The evolution of true vision, the ability to perceive extended space and "distance" in it, meant that the

* William Blake expressed all this much more succinctly in four lines:

To see a world in a grain of sand,
  And a heaven in a wild flower,
Hold infinity in the palm of your hand,
  And eternity in an hour.

creature could interpret and make sense of such contradictory information.

Early human beings may well have faced problems similar to the one above as symbolism evolved. Information received in symbolic codes could contradict perceptual information. The name of a long-dead person, for example, evoked the image of the person's face, the sound of the person's voice, yet the person was nowhere in the perceptual world. We can assume that it was the evolution of symbolic time, the consciousness of an extended "past," which enabled the brain to interpret and make sense of such contradictory experiences. Consciousness of the "past," a symbolic time, was a new phenomenon—a new mental activity different from simply having memories. The latter capacity was common enough in the animal world. The former is symbolic and evolved in humans only.

We can also speculate that it was the confusing and frightening phenomenon of death that prompted the extension of perceivable space into some ill-defined "elsewhere." This probably occurred early in human evolution. Long before the appearance of modern humans (homo sapiens sapiens), earlier human races like the Neanderthals (also of the genus *Homo,* but designated merely as homo sapiens) buried their dead and put food, tools, and weapons in their graves, evidently assuming that these were to be needed sometime somewhere. The making of complex tools, also predating the appearance of modern humans, was probably also an important factor in the development of the concept of an extended future with a concrete shape. The creation of more complex tools obviously required an a priori idea as to how and on what occasion they would be used; they could not have been designed otherwise.

An interesting feature of the evolution of extended time, of the consciousness of past and future, was that it created a mental image of "time" which we feel—quite literally—to possess a kind of physical extension. This mental picture is still very much with us and makes us think of or visualize "time" as if it were something like space. Thus we speak about the "distant" past and the "near" future, or about the "direction" of time. Space was apparently the most useful model which our brain could use when it extended the perceivable "now" into past and future.

The role of symbolic time and space in reconciling conflicting inputs coming from our perceptual and symbolic worlds remained an important one throughout human evolution. When attractive images, important to the individual or to society, contradicted perceived reality, we merely placed them somewhere in extended symbolic time and space—in the past, the future, or outside perceptual range—elsewhere. The tendency to do this is still strong in us. When the real world does not conform to our ideas, we imagine a more convenient world elsewhere in symbolic time and space: in an afterlife, in the past, perhaps in some utopia. This phenomenon is as widespread in the mental life of individuals as it is in societies.

As symbolic time and space became an extended mental counterpart of the perceivable real world, some of the inborn mechanisms of the brain, mechanisms which had evolved in relation to perceivable space, began to operate in relation to extended time and space. There is, for example, an important inborn program in the mammalian nervous system (and in those of many other living creatures) which makes the creature attempt to control certain parts of its environment for its own or, more often, for the group's benefit. This program is called the "territorial instinct." When this genetic program was extended to symbolic time and space, it produced the desire to control the future, the past, and the elsewhere. Thus symbolic activities aiming to influence the future, to make things happen at some other place, or to replay and thus to control the past evolved. We call these activities "magic," and they became as pervasive in human cultures as the territorial programs were in the biological world.

The inborn drive, strong in advanced animals, to activate genetic models and develop learned ones, to look for stimuli was also extended to symbolic time and space. This is what compels us to learn not only about our physical environment but also about our self-made symbolic world. We look for patterns in space, in time, and in the purely symbolic realm of meanings. What we learn about symbols may no longer be relevant to biological survival, but the tendency to learn is built into our brain and it dies out only to the extent that the brain itself dies out. Some keep their minds active by learning differential geometry, others by listening to gossip about their neighbors. Neither is very relevant to survival, but learn we must; our mind cannot stop any more than our heart can.

The same can be said about our "visual mind." The peculiarly human ability, for example, to "see" real, three-dimensional space when looking at marks on a surface was also probably quite irrelevant in the "struggle for survival." At any rate, no animal seems to have developed this particular ability. But drawings, paintings, and sculptures are the best models of visible reality. They can be studied, manipulated, changed; they are easy to make, and the possibilities they offer are inexhaustible. No wonder they had already appeared in the early days of humanity.

Like mythological sense organs, symbolic time and space enlarged the world of the human mind beyond all limits. The analogy may be more than just a turn of phrase. Human symbolism does function in some respects like a sense organ. It can determine the world in which we live the same way each creature's sense organs determine its particular world. The linguists Edward Sapir and Benjamin Lee Whorf suggested long ago that a particular language contributes a great deal to the formation of the model of the world in our mind. "Human beings," wrote Sapir,

> do not live in the objective world alone, nor alone in the world of social activity as ordinarily understood, but are very much at the mercy of the particular language which has become the medium of expression for their society. . . . The fact of the matter is that the "real world" is to a large extent built up on the language habits of the group.

Whorf argued that the symbolic times and spaces of societies are determined by their language and attempted to show that the Hopi Indians in the southwestern United States, for example, have notions of time and space different from those of societies using Indo-European languages.

No consensus exists on these and related matters among philosophers or linguists and none is likely to emerge soon. Other ideas concerning the relations between language and time and space have also been put forward from time to time. About half a century ago, the Canadian linguist R. A. Wilson suggested that the evolution of language itself was generated by the need to express the symbolic time and space of the human mind. Language, for Wilson, is "the completely efficient instrument for the elaboration of the space-time world of free mind." (The latter expression means something similar to what we are calling symbolic space and time.)

Be that as it may it is known that human societies regulated their lives by symbolic times and spaces from prehistoric times on. Holy days, sacred places and directions, working and rest hours, private and family plots, tribal or national boundaries, sanctuaries, plays, musical forms, dances, ritualistic spatial arrangements like those at Stonehenge, statues, magic drawings, ceremonies, were among the symbolic forms in space or time humans used throughout history in trying to make sense of their experiences.

Human symbolism turned out to be a great evolutionary force. It transformed the face of the earth profoundly and it is on its way to influencing and changing the large-scale cosmic environment, as well as the very process of biological evolution itself. It also becomes clear that the more abstract some of our symbols become—the more detached they are from immediate reality—the more powerful they become.

Physics is a case in point. In 1968, for example, when Major William A. Anders radioed back from Apollo 8 on its voyage around the moon that "Isaac Newton is doing most of the driving right now," his message pointed to the awesome power of the symbolic time and space of Newtonian physics. For it was by manipulating the abstract symbols of Newtonian time and space that the journey to the moon was planned down to the last detail. We are still slow at manipulating the symbols of physics—we can still catch a ball faster than calculate its trajectory—but our symbols have already taken us much farther and much faster than DNA-programmed evolution alone ever could.

The even more abstract symbols of modern physics have given us the power to destroy our very species, our past accomplishments and future hopes, our accummulated genetic and extrasomatic records, and our habitats, together with a large number of other species: to destroy, literally, the human role on earth. If such a horror ever comes to pass, it will come as a result of the perceived defense of some human symbols, from the perceived need of protecting symbolic spaces like national territories or boundaries or defending symbolic times like the shape of the future as seen through religious or ideological beliefs.

4

Early Human Cosmologies

(Time and Space Acquire Meanings)

And, as imagination bodies forth
The forms of things unknown, the poet's pen
Turns them to shapes, and gives to airy nothing
A local habitation and a name.

Shakespeare, *A Midsummer Night's Dream*

We would probably be better able to understand ourselves if we knew how the human horizon expanded and how our minds gradually embraced the manifold infinities of time and space. If we could see how certain places and times became more important than others, how the symbolic worlds of the sky evolved, and how certain simple rhythms, words, and figures acquired a literally cosmic significance. But only a mere fraction of this process will ever be known. Only from the last few hundred years do we have relatively satisfactory accounts of the history of ideas. From earlier times we have mostly fragments, often merely translations of incomplete copies of long-vanished originals. And from before the recorded era we have even less information—and none of it direct—as to how early humans thought and felt.

Two examples from prehistoric times come to mind, however. First is the discovery that our long-extinct aunts and uncles, the Neanderthals, not only put food and weapons in the graves of their dead, but also seem to have performed ritual human sacrifice as early as seventy thousand years ago. It is difficult to imagine the staging of such deliberate actions, such emotionally charged ceremonies, without the underpinning of a certain system of beliefs. Such a system would most probably have included a symbolic cosmology of netherworlds, of imaginary pasts and futures: some fairly complex worlds in symbolic time and space created to transcend the here and now. If this is true, then symbolic cosmology preceded anatomically modern humanity since the Neanderthals were anatomically different from homo sapiens sapiens (contemporary humans).*

Second, we can look at the now very famous cave paintings in the Pyrennees—in the Altamira caves in Spain and the Lascaux caves in France. We know the approximate age of these paintings, and, by excavating in their vicinity, we have learned a lot about the everyday life of the people who created them. But the actual role and significance of the paintings themselves still completely eludes us. It may be a mistake, for example, to assume that the paintings represent everyday occurrences. A visiting archaeologist from another planet, knowing nothing about the European Middle Ages except its paintings and sculptures with their preponderance of crucifixes and martyred saints, could logically conclude that the most prevalent activity of the period was torture. And while there was a great deal of torture in the Middle Ages (perhaps almost as much as in the twentieth century), the conclusion of the visiting scholar would still not really reflect the daily realities of medieval life.

The only conclusion we can draw, therefore, from looking at the cave paintings is that about fifteen thousand years ago a group of paleolithic hunter-gatherers were able to depict spatial forms and relations and the motion of animals with considerable skill and authority. What this shows is that these people already possessed a high degree of sophisticated intuitive understanding of their visual world and that they knew how to re-create this world convincingly in a symbolic two-dimensional representation.

So it seems altogether safer to start with some symbolic cosmologies current at the time of the first written records. However

* They had, for example, larger—although not necessarily better—brains corresponding to their larger bodies.

woefully inadequate they are for our purposes, it is in these records that we find descriptions of the earliest known symbolic attempts to make sense of the world: the sweeping, colorful theories which we now call mythologies. Until the birth of modern science, these theories were our only conceptual cosmologies, and although science and mythology are different undertakings, they also have some features and functions in common. Both offer symbolic representations of the world; both look for unity in its apparent diversity; both look for order in its apparent disorder; and both imagine and explain the world in symbolic time and space.

In the early mythological cosmologies, time and space were not generalized, independent concepts but, like most other mental constructions, were human symbols carrying emotional significance. In early—and in most traditional—societies, the mammalian view of the world, the inborn cosmology of the human senses, became so intertwined with the culturally evolved symbolic world that the distinction between the two was blurred. For ancient humans, therefore, it was difficult to separate a symbol from the thing it represented. This particular problem, the identification of a thing with its symbol, persists to some extent to this day, but, in the thinking of ancient humans, this "coalescence" of a symbol with what it represented was very pronounced. Words, dreams, and even artifacts could assume the role of the thing—or being—they represented. Thus the statue of a god became the god and was worshipped. The name of a god became holy because the god was holy and the name itself was therefore treated with the highest respect. "Whatever is capable of affecting mind, feeling or will has thereby established its undoubted reality," summed up the late Henri Frankfort in a classic study of the thinking of ancient humans.

It is possible that this intermingling of symbols and reality simplified the mental world of the newly emerging civilized societies and made adaptation to their increasingly complex social world easier. And it was perhaps for the same reason that the distinctions between objective and subjective, between individual self and society, between causality and coincidence, dreams and real events, the living and the dead, were equally vague in these societies. Therefore it is no surprise that there was little difference between perceptual space and the symbolic spaces of the imagination, and that the abstract concept of space as we know it today did not figure in the thinking of ancient humans at all.

The concept of space, according to Frankfort, was not abstracted from the experience of space. Instead of thinking in terms of abstract space, the ancient mind thought of a concrete place. "The spatial concepts," writes Frankfort, "are concrete orientations; they refer to localities which have an emotional color; they may be familiar or alien, hostile or friendly." The same was true of cosmic events. Ancient humans felt the need to endow space with particular emotional meanings: "Day and night give to east and west a correlation with life and death." Fulfilling a similar emotional need, the Babylonians used the observational data of their highly evolved astronomy to create an imaginary, but "extensive system of correlations between heavenly bodies and events in the sky and earthly localities."* Thus, concludes Frankfort, primitive thinking "may succeed no less than modern thought in establishing a co-ordinated spatial system; but the system is determined not by objective measurements, but by an emotional recognition of values."

The emotional and symbolic significance of places or directions was determined by the particular dominant mythology. Carrying the heaviest emotional weight were the most important early symbolic spaces: the sacred places and directions. The dominant mythology located these and determined the degree of their sacredness. These ideas may not always seem very logical to us. Frankfort points out, for example, that the sacred inside of an Egyptian temple derived its sacredness from the belief that it was at that particular place that the creation of the universe took place. However, the same claim could be made by other temples, and all such claims were accepted as truth by the faithful. Since all these places were endowed with the right symbolic significance the claims were perceived as being consistent with each other. They did not violate either the conceptual or the perceptual logic of their societies.

In addition, these claims answered an important societal need: the need to believe that the world was created on Egyptian soil. In

* Astrology still survives among—to quote the historian Sir William C. Dampier—"the uneducated in all classes."

most mythologies, the entire creation of the universe was centered around a particular tribe or nation. The symbolic space in which the universe unfolded was therefore essentially the space of the community—a tribal space. This particular intuition, by its nature, limited the extension of the imagined symbolic spaces of ancient humans. Space ended where the community's power or interest ended. According to this view, there was the territory of the tribe or the nation and there was the "unknown and indeterminate space that surrounds it," says the historian of religions Mircea Eliade. The former was perceived as sacred, favored by the gods; it was friendly and unchanging, lawful and ordered. The latter, in contrast, was structureless and frightening, chaotic and dangerous. It seems, in short, that in most ancient civilizations, the property of the world that we now call space was thought and felt to be something subjective, a tribal property, limited and closed.

As to the sensory perception of space, very little is known directly. It is usually assumed, however, that the surviving remnants of the plastic arts do, to some extent, reflect a people's perception of their spatial environment. Buildings, paintings, and sculptures document the spatial intuitions of their creators and of the societies which produced them. The surviving artworks of a civilization are therefore important not only for their inherent artistic-aesthetic-emotional values and not only for what—in conjunction with written records—they tell us about the ideas and beliefs of society, but also because they were created by a mental process in which the artist reconstructed the perceived world in a symbolic space. This space reflects the artist's perception of the world and therefore reflects the emphasis on and significance of that world's various visual and tactile properties.

If we take a look at Egyptian block statues, we can see a clear example of how an artist's or society's perception of space can dominate artistic expression. The first known examples of these characteristically Egyptian creations (although similar creations can be found in the remnants of pre-Columbian civilizations) are from about 2000 B.C., and they were still being made towards the end of Egyptian civilization during the last few centuries before Christ. Figure 4-1 shows a typical Egyptian block statue. The sculptors of these powerful figures have achieved—and the combination is a powerful one aesthetically—the realistic depiction of

Figure 4-1

a human body within an abstract and regular geometric form; their work reveals the imposition of a rigid spatial order on human existence. The visual and tactile space that the piece seems to impose on its viewer is closed, inward looking, and static. The figure is deliberately separated from its surroundings by sharp, well-defined boundaries. Everything important refers to the inside; the work contains almost no reference to the outside.

A comparison with a Greek sculpture made some fifteen hundred years later (Figure 4-2) makes the point even clearer. The Greek figure, in perfect contrast with the Egyptian, is actively expanding into its surroundings. It creates an open visual space with its own dynamics and is quite ready to move into that space. The sculpture radiates a feeling of confident activity and a kind of actualized power which is altogether different from the sense of hidden potential which emanates from the Egyptian figure. One could, of course, easily find early Egyptian statues that are less static and Greek ones that are much more closed. Nevertheless

Figure 4-2

these examples are not misleading. They define the different qualities of perceptual space in two different civilizations.

When we compare Egyptian and Greek architecture, we receive similarly different impressions. To achieve monumentality, the typical Egyptian temple employs a solid, compact mass and makes little use of open space. The Greeks, on the other hand, had a much "airier" approach to building. Rather than avoid large expanses of empty open space, they made them an aspect of the grandeur of the building.

Early Egyptian and much other archaic painting offers a similar impression of a closed, two-dimensional space. The viewer is not invited to "walk into" the paintings. There is no visual depth—most ancient paintings are as flat to the eye as to the touch. The subject matter of these early paintings, which is mostly mythopolitical, reveals the same lack of distinction between the real and imaginary that characterizes ancient conceptual thinking. To give an example: Since there was no clear conceptual distinction between the living and the dead—who, although somewhere else, continued to interact with the living—so there was no clear separation between real perceptual space and the imagined space of the dead, the netherworld. Like words, paintings and other artworks also substituted—in the ancient mind—for what they represented. While the statue of the god was worshipped, the figure of an enemy in a relief was spat upon. The relative size of objects and persons, to take another example, was not determined objectively by either tactile or visually apparent size but by their relative symbolic importance. Thus, in Figure 4-3, the Egyptian nobleman is much larger than the serfs who serve as his boatmen. This manner of depicting people (which, incidentally, was typical of many civilizations including that of medieval Europe) probably reflected the way people actually saw the world.* For until modern science became influential, it was a generally accepted truth that the bodies of royalty and nobility were different (more beautiful, larger) than those of commoners or serfs. And the ancient Egyptians—just like us—saw what their minds prepared them to see.†

As with space, the concept of time in ancient thinking was not separated from the actual experience of time, which was deter-

* Painters in Western civilization abandoned this idea some time ago. But this same rank-space symbolism still survives elsewhere. Thus all governments and other large organizations allocate more office space to higher-ranking personnel. The Federal Property Management Regulations in the United States, for example, permit a secretary to have 60 square feet and a middle management supervisor 220. The ratio is not very different from that in Figure 4-3 and, of course, usually has nothing to do with the actual space needed for performing their respective duties.

† By this I do not mean to imply that the ancient Egyptians saw their daily surroundings any differently than we do. It is a remarkable human capacity to live simultaneously in the everyday perceived world and in symbolic worlds. People in most civilizations, for example, believed that a "better world" awaited them after death. Yet few people actually welcomed their own or their beloved ones' deaths. The contradiction between the biological instinct for survival and their symbolic beliefs seldom caused serious discomfort to most people.

Figure 4-3

mined by the rhythms and periodicities of nature as reflected in individual and social life. These rhythms and cycles were as fundamental to human as to primordial life.

Since their livelihoods often depended on a knowledge of the life cycles of plants and the migrations of hunted animals, even early hunter-gatherer societies were aware of the periodicities of time. With the evolution of agriculture, the ability to predict environmental events became even more important. Since the human body does not possess manifest circannual or photoperiodic clocks to signal seasonal changes, societies had to invent the cultural equivalent: the calendar. The calendar was the first symbolic construction that regulated social behavior by keeping track of time. Similarly, the importance of the day-night alternation in individual and social life also led to the evolution of symbolic clocks, cultural equivalents of the circadian biological ones. But the need for such clocks was not as pressing as the need for a calendar, both because people do have internal circadian clocks and because the simplest external one, the sun, was almost always readily available. Early humans could probably tell the time of the day from the position of the sun and could thus regulate their daily activities long before the appearance of any artificial clock.

All this, however, was more a reflection of experience than of any abstract concept of time. In ancient thinking, time, like space, was concrete, defined by and related to events, colored by emotions, and laden with symbolic significance. But the overwhelming property of time for ancient humans was its cyclic, repetitious character. In the emerging world of agricultural societies, the periodic changes in nature were of the greatest importance, and therefore the first sophisticated human cosmologies viewed time as an infinite sequence of recurring events. As Mircea Eliade writes, "The cosmic cycle is conceived as the indefinite repetition of the same rhythms: birth, death and rebirth."

This periodic, circular view of time, with its roots in the mechanics of the solar system and with its manifestations in the day-night and seasonal cycles, was a superbly useful symbolic model. It allowed the world to be predictable and stable; it established order and reduced uncertainty. Circular time was also reversible; events could be relived again and again. "Every New Year is a resumption of time from the beginning; that is a repetition of cosmogony," points out Eliade. Repetitive time even made the idea of death tolerable by promising rebirth.

Its adaptive advantages and its essential congruence with the workings of the environment explain why this particular symbolic time evolved independently in so many civilizations in the course of history.* And wherever it evolved, this circular view of time

* In Indian cosmology, the world is destroyed and re-created about every four billion years. During this period there are also other recurring great events of shorter duration. The Babylonians believed in a fundamental period, the Great Year, which lasted four hundred and thirty-two thousand years and after which the world was destroyed by a universal flood. The idea of periodic, recurrent time was dominant in the ideas of many—but not all—Greek thinkers. Plato, for example, seems to have believed that time itself was created by the periodic

gave rise to characteristic cosmologies which share many similar features. Since in these cosmologies the future was merely the repetition of past events, whatever happened in the past would happen again in the future. Therefore, the distinction between past and future was unimportant and irrelevant, and the passing of time was viewed as an illusion. In this sense, time was as closed for early humans as space. The entire world, in the ancient imagination, was a small, enclosed space in which events were repeated for ever and ever in rhythmically recurring time.

But although it offered security, this cyclical worldview was also crushingly static. Novelty or change of any kind was frightening because it threatened the built-in stability provided by the ever-recurring cycles. Many of the ideas (such as "history," "evolution," or "progress") which later became very important to us could never have developed in societies which believed strongly and monolithically in rhythmic cosmologies. A strong belief in circular time will, usually, give rise to a fatalistic worldview which stifles curiosity and discourages initiative. Efforts to improve society, for example, seem quite pointless if everything is assumed eventually to return to the same state.

The pull of cyclic time remained a strong one in most civilizations throughout history. It seems, in fact, that only Western civilization managed to free its collective imagination—to some extent and for the time being at any rate—from the trap of cyclic time. And it was also Western civilization only which liberated its spatial conceptions from the prison of the self-centered, closed tribal spaces. It was, of course, due largely to the development of the natural sciences that our present ideas of progressive and irreversible time and homogenous, boundless space were established. But the roots of these ideas go way back in history—right back to the founding parents of Western civilization, Hellas and Israel.

When historians write about these ancient cultures, Hellas is usually credited with, among other things, inventing coherent, rational thinking across a wide range of issues of human interest,

---

motions of the planets and that without this motion there would not be anything called "the moving image of eternity"—his description of the passing of time.

and with introducing realism and visual logic in the arts. Israel, on the other hand, is said to have established the groundwork for ethics and social justice. But another equally important aspect of the legacies of these ancient peoples is that the Greeks of antiquity were the first to create—through geometry, art, astronomy, and philosophy—symbolic spaces which were free of supernatural values and superstitious symbols. These spaces were characterized instead by logically and perceptually objective properties. The ancient Jews, on the other hand, invented a cosmology in which the passage of time was no longer circular, and evolved a corresponding worldview in which the concept of the future became important.

The cosmology of the ancient Jews is described in the Bible. It starts with a creation legend, recounted in the book of Genesis. This legend had a great many thematic similarities to other, earlier Near Eastern myths. But it also possessed several unique characteristics, some of which influenced Western thinking throughout history. One such characteristic was that in the biblical cosmology neither the sun nor the moon was perceived as having any kind of supernatural or holy quality. Both were viewed as mere objects, created by the Lord to serve humankind by providing light for the day and for night respectively. They came into existence as the "handiwork" and moved at the pleasure of an omnipotent God. Both could be used to signal or to measure the passing of time (Jews, like other Near Eastern peoples, used a lunar calendar), but in the biblical cosmology they did not "create" time nor did they influence the fate of the world in any important way, as they did in many other cosmologies.

The way the world unfolds, according to Jewish cosmology, is determined solely by a single, omnipotent God. His will is absolutely free and He is therefore under no compulsion ever to repeat events. And in fact, in the biblical story at least, God displays no inclination to do so. On the contrary. The creation of the world was, in the Bible, a unique event and not just one rebirth in an unending growth-decay cycle.* This novel way of viewing the

* A characteristic example of this new perception of the unfolding of events is the following: The biblical story of the Deluge, like many other biblical stories, had its origins in much earlier Mesopotamian legends. In these earlier legends,

world was very different from most ancient cosmologies, and it was based on this view that a new kind of symbolic time began to take shape.

Since this new "time" was not bound to the cyclical mechanics of the solar system, it was able to slowly evolve toward a "time" that was a progressive, linear, nonreturning flow. In the framework of this new symbolic time, the future became different from the past—and not only different, but also uncertain and open-ended. God's will alone could change it, and God's will could be influenced by human behavior. The Jews, in other words, did not project the periodicity of the solar system or the birth-death-rebirth cycles of agriculture onto the cosmos. Correspondingly, the early Jews did not believe in the survival or rebirth of the body or soul. Only much later was this idea taken over from other religions. The original biblical cosmology was an intuitive and original attempt to describe and make sense of the world in a linear, noncyclic, open-ended, symbolic, time.* In this emphatically nonfatalistic framework, furthermore, it was no longer credible that the future was determined by the motion of the stars or that it could be predicted by looking at the intestines or livers of animals. Thus this view of progressive time led to a generally hostile attitude toward astrology and many other contemporary forms of magic.†

---

however, the destruction of the world by water was a recurrent event which had taken place innumerable times in the past and would occur regularly in the future. Genesis kept the flood story but transformed it into a unique event. In this view, furthermore, the stability of the world was assured precisely by the uniqueness of the event, i.e., by the heavenly promise that the Deluge will never happen again.

* The attempt was almost certainly original, but it may not have been the first in all respects. The Zoroastrian religion also views time as progressive—but not open-ended since the ultimate victory of good over evil is part of the original religious doctrine. This religion, however, has not played a significant role in Western thought.

† The German sociologist Max Weber sees this attitude as one of the historic components which led to the evolution of the Western rationalism underlying the sciences, arts, and social life of our civilization. It seems that this "hostility to magic"—as Weber calls it—was a natural consequence of the view of time as progressive, open-ended, and nonpredictable instead of predetermined and circular.

Another remarkable feature of the biblical cosmology was that it viewed time as a much more significant dimension of life and of the universe than space. This view was already apparent in the creation legend. The Genesis story does not mention and displays no interest in *where* the world was created.* Instead it classifies the processes of creation by giving their exact temporal order. Consequently, the act of creation did not define any sacred spot, site or region. But the temporal organization of the very same primordial act did define an all-important sacred time: the Sabbath. In its further evolution, the ancient Jewish cosmology displayed an intense hostility toward any symbolic spatial representation of God. The making of idols was strictly forbidden. The only symbols allowed to represent God were words: His name was holy; His words were holy. He had no spatial form and therefore could not be seen or touched. But He could be heard. He often talked to His servants, but mostly from nowhere in particular.

This purely religious, time-oriented cosmology stands in fascinating contrast to the dominant ideas of the Greek civilization. Greece, of course, produced so many and such diverse great philosophers that it is not easy to speak about a "Greek" philosophical view of time. The germs of virtually every rational idea about time figured in the thinking of some Greek philosopher. Nevertheless the overall thrust of the Greek view of the world was always toward eliminating the dimension of time, toward reaching the "eternal"—that is, the timeless—truths about the workings of the world. It is instructive to compare biblical accounts of the creation of the world with a sophisticated Greek cosmogony such as that described in Plato's *Timaeus.* The Bible describes the creation as a temporally unfolding event without exhibiting the slightest interest in its spatial whereabouts. In Plato, the creation of the world had virtually no temporal dimension. The time-order of the process, the temporal sequence of events, was as irrelevant to him as the spatial order was to the Jewish cosmogonists. The essence of the act of creation was for him the realization of perfect, unchanging geometrical mathematical laws through material exis-

* Genesis seems to have been written by several authors living at different times. Its first two chapters, for example, already contain two non-identical creation legends. This, however, is irrelevant for our purposes as we are interested in the main thrust of the biblical cosmogony only.

tence. But, in his view, the very fact of material existence already debased these perfect laws and made them imperfect, changing, and of a lower order. Essential reality lay therefore in a realm of eternal, perfect laws, and whatever we might learn about this eternal realm with our senses was a mere likeness. Even time itself was nothing in Plato's view but the "moving image of [this] eternity." Space, in contrast, was not just an image. It was very real, so real that it existed before the universe itself.

In emphasizing the timeless character of the laws governing the world, Plato, of course, anticipated certain developments in modern science and, in particular, in modern theoretical-mathematical physics. Nevertheless it was this characteristic disdain for every kind of change, for all things temporary, which prevented the Greek thinkers from ever coming to grips with the basic problem of all natural science: the problem of simple motion. For all their awe-inspiring brilliance, not even the brightest of them could come close to comprehending the idea of temporal evolution. Thus Zeno of Elea, who lived a generation before Plato, considered ordinary motion nothing but a sensory illusion and devised ingenious examples (the famous paradoxes) to prove this assertion. Aristotle, whose fertile mind made lasting contributions to everything from logic to biology, thought that since the spatial appearance of a circle was more "perfect" than that of a straight line, circular motion was simpler than rectilinear!

Rhythmic and therefore essentially unchanging history was a satisfactory reflection of the timeless character of the world. And many influential Greek philosophers, Plato, Aristotle, Pythagoras, Heraclitus, and later the philosophers of the Stoic school among them, believed in cyclic time.* But, in contrast to many other civilizations, these views did not influence Greek metaphysical thinking very deeply, precisely because time was not an important

* On the other hand, the only example in the Hebrew Bible of a cyclic view of time is in Ecclesiastes. ("The thing that hath been, it is that which shall be; and that which is done is that which shall be done; and there is no new thing under the sun.") There seems to be ample evidence that this book was written in the third century B.C. by thoroughly Hellenized Jews in Greek-ruled Judea. The book is full of Greek ideas. It starts, to give an example, by enumerating the four elements (earth, fire, air, water) which formed the material world according to Greek natural philosophy. And for the first time in the Bible, the ascending of souls to the heavens, another Greek idea, appears in this book.

component in their worldview and, therefore, it did not matter very much whether it was periodic or not. For this reason, astrology was not very important in Greek civilization, and the Greeks were able to write the first rational histories without bothering too much with recurrent time.

The realm in which the Greeks' quest for eternal timeless truth succeeded brilliantly was symbolic space. It was in the development of geometry and its associated mathematics, in the generation of philosophical ideas about space, in astronomy, and in the visual arts that the Greek genius was most fertile. Thus the boldest of all conceptual symbolic spaces—limitless, infinite, homogenous, empty space—already appears in Greek thinking. It made its first appearance, it seems, in the writings of Demokritus. Demokritus (apparently following the ideas of his teacher Leucippus) thought that all matter was composed of invisible, indestructible small particles—atoms—which move in infinite space. Their movements, collisions, and alignments make the material world. Later it was Epicurus and the Roman poet-philosopher Lucretius Carus who developed and propagated the idea of objective, infinite space as a receptacle for bodies—the symbolic space of the classical physics of Galileo and Newton a millennium and a half later.

Far more influential than their philosophical concepts were, however, the earliest Greek attempts to find abstract laws to describe spatial forms and ways to prove these laws in a general way. There is some evidence that it was Thales (about 630–545 B.C.) from the city of Miletus in Asia Minor who first gave an actual proof of a geometrical theorem. If this is true, then Thales not only discovered a law of geometry but was also the first who consciously applied the deductive method to an abstract problem. This would make him an epochal figure not only in mathematics but in the entire history of thought. Another even more influential figure was Pythagoras, who was active around the beginning of the fifth century B.C. Bertrand Russell calls him "intellectually one of the most important men that ever lived, both when he was wise and when he was unwise." He stands indeed like a great divide between ancient-mythical and modern-rational thinking, embracing both in his colorful, imaginative theories. He was born on the island of Samos but was active in the Greek city of Croton in southern Italy. There he founded an esoteric religious-mystical community where, however, some excellent mathematics was also cultivated.

True to the Greek genius, the Pythagoreans imagined numbers as actually having spatial extensions and shapes. For example, even numbers which could be decomposed to unequal factors were called "oblong" and were imagined to look like this:

$2 = 2 \times 1$ $6 = 2 \times 3$ $8 = 2 \times 4$

Not all numbers were oblong; there were also "square numbers":

$4 = 2 \times 2$ $9 = 3 \times 3$

Odd numbers, on the other hand, had shapes like this:

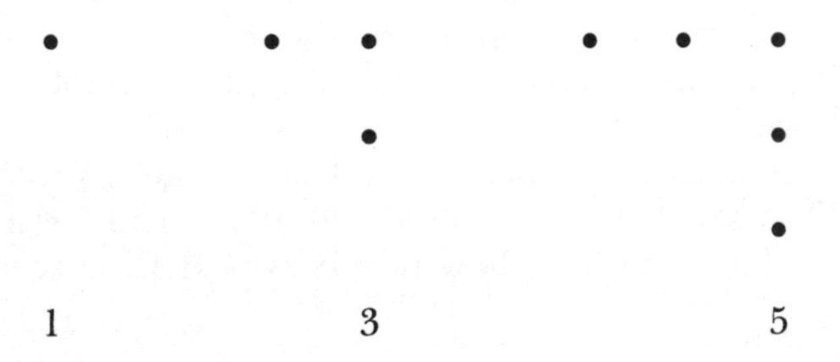

The Pythagoreans then put together the forms of consecutive odd numbers and always got the form of a square number.

Equally obviously they also got all square numbers:

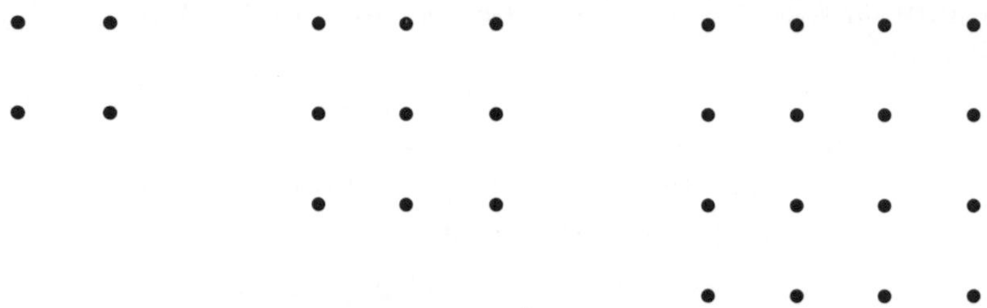

• •	• • •	• • • •
• •	• • •	• • • •
	• • •	• • • •
		• • • •
$1 + 3$	$1 + 3 + 5$	$1 + 3 + 5 + 7$
$= 2 \times 2$	$= 3 \times 3$	$= 4 \times 4$
$= 4$	$= 9$	$= 16$

This is a typical Pythagorean proof of an abstract theorem in number theory.* The Pythagoreans did not need abstract concepts or algebra to prove this theorem; they saw that it was always true—no matter how far you went or how large the numbers in question—by comprehending the spatial structures.

They made a number of similar discoveries with the help of their figures. The realization that they could think up such abstract and timeless truths and could actually prove them must have been intoxicating. No group of people before the Pythagoreans could possibly have felt the power of human intellect in quite the same way. It is understandable that they believed themselves to be on the threshold of finding the basic design of the universe through numbers, through arithmetic alone.

This belief was further reinforced by their discovery—this time not theoretical but experimental—that the pitch of a sound emitted by a vibrating string is governed by simple numerical ratios. If, for example, the lengths of the strings were in the ratio 6 : 4 : 3, the sounds produced would be a note, its fifth, and its octave. They saw how the senses could discover numerical laws in the observable world and how numbers underline reality. They

* In more modern notation:

$$1 + 3 + 5 + \ldots + (2n - 1) = n^2$$

for all natural n's: $n = 1, 2, 3$

jumped to the general conclusion that the whole universe, including life, society, and ethics, could be explained by numerical laws, and they founded even their moral beliefs on this assumption.* At the time when much of human thinking was still largely mythical and magical, the Pythagoreans saw both rational and magical power in numbers.

Ironically, their worldview collapsed when they made their most famous discovery: the Pythagorean theorem of right triangles. It seems that the Egyptians had known a long time before that a triangle with sides 3, 4, and 5 must have a right angle and that they used this knowledge (allegedly secret, known only to a few priests of sufficiently high rank) to form right angles in buildings. Pythagoras apparently learned this rule during his travels to Egypt. He may have noted that $3^2 + 4^2 = 5^2$, and, perhaps taking this as a hint, he discovered and proved the general theorem: that the sum of the squares on the sides adjoining the right angle is equal to the square of the third side, the hypotenuse.

But when the Pythagoreans tried to use this theorem to find the length of the diagonal of a square whose sides were of a unit length, they found to their dismay that no matter how small they made their units, there was no number which could express this measurement. The problem was not that the length of the diagonal could not be expressed by integers, but that it could not be expressed by fractions either. The measure could not be expressed by any number known to them at the time. The length of this diagonal is, according to the Pythagorean theorem, $\sqrt{2}$ times the unit length, but $\sqrt{2}$ is neither an integer nor a fraction. It is an endless decimal: 1.4142. . . . Such numbers are now called irrational; the name still reflects the shock that the Pythagoreans felt upon making this discovery.

Thus the idea that all important properties of the world could

* They thought, for example, that the distance between the heavenly bodies corresponded to the lengths of strings that sounded different musical notes. They also thought that the heavenly bodies were carried by spheres which made music while turning. This was the origin of the metaphor of the "music of the spheres." A sample of their mathematical morality: Even numbers were female, odd ones male. No need to add that they found even numbers "bad" and odd numbers "good."

be expressed by numbers alone received a serious blow when it turned out that not even the length of the diagonal of a simple square would conform to it—at least not using any numbers they knew about. The reaction of the Pythagoreans was, at first, the same as that of all fanatics when confronted with facts that do not fit in with their beliefs: They tried to suppress the discovery. They did not succeed. And in the end, their ideas influenced many thinkers from Plato on.

There were, however, important Greek thinkers who were not quite so mesmerized by numbers and geometrical symmetries. While Plato advocated that the senses could not yield "real" knowledge, others would not accept such a separation of the space of sensory experience and the space we think about.* Eudoxos, a younger contemporary of Plato, created a complicated and quite wrong, but nevertheless rational theory to account for the real motion of the sun, the moon, and the planets. This was an early example of the then new Greek idea that one could give an *explanation* for observed phenomena on a rational basis; in other words, that one could construct *rational theories* in symbolic time and space to account for what was observed in perceptual time and space. The Babylonians had made excellent long-term astronomical observations, had collected a vast amount of data and used it for practical and religious purposes. But it had never occurred to them to separate time and space from religion, and make rational, nonmythological explanations for celestial phenomena.

Plato's great pupil Aristotle (384–322 B.C.) also criticized his master's teachings and emphasized the importance of observations. He was perhaps the first who connected observations with explanations throughout a wide range of phenomena. Aristotle observed both plants and animals; his ideas about biology were new, and many turned out to have lasting value. He was the first to realize that the fundamental problem in understanding how an individual organism develops lies in explaining how the information governing its growth could be contained inside the seeds

* Plato believed to such an extent in the illusory quality of sensory information that he suggested that the pursuit of astronomy be conducted without paying too much attention to the sky: "Astronomy, then, like geometry, we shall pursue by the help of problems, and leave the starry heavens alone."

from which it grows. This problem was not solved until the discovery of the structure of DNA in 1952.*

These two ideas, the importance of the senses in observing the real world and the quest for symmetry and spatial order, were argued over by philosophers, but were harmoniously incorporated into the Greek visual arts. This made the art of the Greeks as unique among ancient cultures as their philosophy and mathematics. Greek art was, to start with, the first "free" art anywhere in history—free in the sense that its purpose was, at least in part, aesthetic and not religious or political. It was also far more realistic than anything created before because the Greek artists were the first who thought it important to observe the human anatomy as well as the dynamics and equilibrium of the human body. I have already mentioned the sculpture in Figure 4-2. It is dated from the beginning of what is now called the Classical or Athenian era.† It was made around 450 B.C. and is one of the very few surviving original Greek bronzes.‡ It depicts either the chief god Zeus or Poseidon, the god of the sea; opinions differ. It does not, of course, really matter which one it was intended to depict. What matters is what we see: a three-dimensional image of a nude male in a pose which indicates that he is ready to hurl something. We also see that this man is in perfect dynamic equilibrium and will not trip after he lets the missile in his right hand—which has disappeared sometime during the centuries—go. The figure and the pose are expressive and beautiful, and the piece stands as a monument to a detailed study of human appearance and anatomy. The limbs, underlying bone structure, and muscles have all been depicted with great accuracy. No such sculpture could possibly have been produced by any other civilization of this or an earlier epoch. In fact, none of the Near Eastern civilizations produced freestanding three-dimensional sculpture depicting mo-

*To emphasize the importance of this observation, the physicist-biologist Max Delbrück of the California Institute of Technology called Aristotle, only half in jest, the co-discoverer of DNA.

† It is customary to divide Greek civilization from the eighth (Homer's) century B.C. on into three different eras. Fusing the periods and nomenclatures of art history and the history of science, we can call them the Early-Archaic (end of the eighth century to about 480 B.C.), the Classical-Athenian (480–320 B.C.), and the Hellenistic-Alexandrian (320–100 B.C.).

‡ It is now in the National Museum in Athens.

tion realistically.* Nor, until the Renaissance, was any produced in Europe! This history alone shows how bold a spatial imagination was required to form such a creation.

Greek art also reflected the quest for symmetry and spatial order. For the Greek imagination, beauty, like truth, was thought to result from a harmonious order. In fact, in Greek thinking, it was not quite possible to separate rational and aesthetic (and often even moral) aims and aspects.

Consider the female figure in Figure 4-4, which was conceived close to the end of the Classical period in about 350 B.C. It is a Roman copy of the lost original by Praxiteles, and it is the most celebrated artwork of the fourth century B.C. It depicts Aphrodite, the goddess of love. It seems that the head, hands, and feet were added later, so it is the torso and the legs only which may be similar to the original. We can guess with some confidence that the sculptor's aim—conscious or not—was not to re-create the image of a real woman. He wanted to create, instead, an ideal female figure with perfect proportions, a figure which would "move" in the most graceful way imaginable. The statue is a visual embodiment of the Greek idea that the aim of an artist is to create harmony and beauty in a spatial form, to show the ideal perfection of the human body as it shines through the wrinkles and disproportions of the real thing. To put their work on what they felt to be an even more secure basis, the Greek artists of the Classical period used—and believed in the importance of—mathematically exact proportions. An artistic treatise by the sculptor Polycleitos (lost, and known only through quotations from it by the celebrated second-century Roman physician Galen) emphasized that "beauty consists in the proportions . . . of finger to finger . . . fingers to palm and wrist and then the forearm, the forearm to the upper arm. . . ." Yet in spite of such principles, these works are not mathematical constructions. They are real art, magically capturing life and motion and evoking feelings which perfect geometrical figures could never evoke.

* The emphasis in this sentence is on the word "realistically." Motion was symbolized in the artworks of many ancient civilizations. Walking, for example, was symbolized by placing one foot ahead of the other. The body as a whole, however, remained rigid and static.

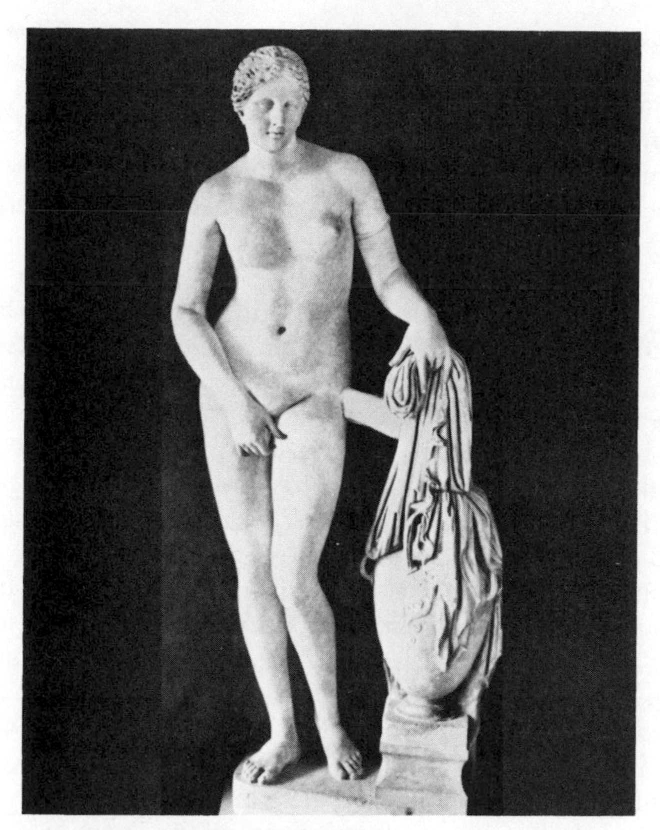

Figure 4-4

The quest for formal perfection was very much a part of the spirit of this extraordinary period. Just a few decades after the end of the era we call the Classical, geometry also reached a perfect and in some respects final form in the works of Euclid (about 330–260 B.C.). In a book written about 300 B.C. and entitled *Elements,* Euclid not only summarized virtually everything that was known in mathematics and geometry at the time, but did this in a logically consistent and self-contained form. No other scholarly work has ever been admired so greatly for such a long time. Nearly two thousand years after its appearance, Newton took Euclid's book as a model for his own work on mechanics, and Spinoza did the same for his work on ethics. Nor did the book's influence diminish with the later blossoming of science and mathematics. Albert Einstein and Bertrand Russell both wrote in their autobiographies that their early acquaintance with *Elements* was one of the decisive factors in steering their interests toward the mathematical sciences. In fact, ever since the Middle Ages, the whole of Europe has learned geometry from this one book. Therefore it is hardly an exaggeration to say that this work was the most influential of all mental creations in shaping the spatial thinking of Western civilization.*

Chronologically, Euclid's creative life belonged to the last stage of Greek civilization. He was educated in Athens but later moved to Alexandria, the cultural and intellectual center of the Hellenistic era. In Alexandria he became the head of the mathematics section of the famous "Museum," often called the world's first university–research institute–library complex. There he wrote all his works. In *Elements,* he attempted to put geometry into an internally consistent, self-sufficient theoretical form. In order to do so he first had to overcome the perennial problem of "definitions." When one tries to define anything, the definition must consist of words; these words then also will have to be defined by new words and so on ad infinitum. Euclid avoided this problem by starting with axioms: statements which are supposed to be so self-evident that they can be accepted by anybody without further inquiry. Then from a few carefully chosen axioms, he proceeded by straightforward logical arguments to derive simple geometrical theorems, and from those, successively more complicated ones. In the end, he could derive from a few axioms all the important results of Greek geometry. Many of his results are well known to us from high school geometry.

Euclid's geometrical axioms and starting definitions indeed looked indisputably and obviously self-evident. Samples: "It is possible to draw a straight line from any point to any point," or "All right angles are equal," or "Things which are equal to the same thing are also equal to one another." Only the so-called fifth axiom sounded somewhat complicated. Stated in more modern language, it is essentially this: If on a plane there are a straight

* In spite of all this, Euclid has never been considered among the most original Greek mathematicians. Neither while he lived nor later was he ever put in the class of Pythagoras, Apollonius of Perga (about 260–200 B.C.), or the greatest of all, Archimedes of Syracuse (287–212 B.C.). Euclid was mainly admired for his ability to synthesize diverse results and connect them logically.

line and a point not on this line, then one can draw one and only one straight line through the point which will be parallel to and thus never intersect the line. This, sometimes called the "parallel axiom," also seems quite obvious to us. Nevertheless its relative complexity aroused the suspicions of mathematicians from early on, and, as we shall see later, the arguments about the logical necessity of this axiom led, in the nineteenth century, to the creation of non-Euclidian geometries and these, in turn, prepared the ground for the twentieth-century abandonment of classical space.

But there were perceptual as well as logical problems with the fifth axiom. That it seems obvious to us does not mean much since we have been brought up and educated in a Euclidian environment with a Euclidean visual logic. But how could this axiom be so self-evident to Euclid and his contemporaries when what it said actually contradicted common visual experience? Our visual apparatus is such that we never see parallel lines extending far away. We see them (like rails, or lines of trees on both sides of a highway) coming together at a distance. How then could this axiom be considered just as self-evident as, say, the one about the equality of all right angles? To explain this and some other features of the Greek spatial imagination, it has often been suggested that in the Greek intuition of space, tactile perception had a greater importance than it has in ours. What do we mean by this? There is some experimental evidence that in our civilization the spatial intuition is dominated by vision. If, for example, vision and touch are given contradictory information in psychological experiments, the subjects believe their eyes before they believe the receptors in their skin. Visual information, in other words, dominates tactile. But this may well be an acquired characteristic of people in a visually active civilization. It is therefore quite possible that in the Greek perception of space, touch had a greater importance than in ours. Touch, of course, is a more active and immediate sense than vision; it makes an impact on its object while vision transmits coded information and has no influence on what is seen.

If the Greeks indeed relied, on the whole, more on their tactile perceptions than we do, this circumstance might help to explain several curious features of their spatial sense. First of all, the fifth axiom becomes obvious since touch never feels parallels coming together. Then there is the Greek discomfort with the problems of motion which I mentioned earlier. Touch, of course, is not well suited for exploring the phenomenon of motion since it cannot easily integrate temporal events. Zeno's paradoxes, which aim at explaining away physical motion as a visual illusion, make more sense in a civilization where touch is the most important source of information. It is also quite natural to imagine that the figures by which the Pythagoreans represented numbers were actually realized by arranging pebbles, and that the exploration of numerical relations was carried out, at first, mainly by touch. Or, to take yet another example, consider Plato's teaching that paintings were always "far removed from truth," while geometry was all important. Geometry can, of course, be taught even to a blind person using touch only. Paintings, however, are irrelevant to the touch and have less meaning for one to whom seeing is less important. It is also to the sense of touch only that a circle, which can be felt in its entirety, might seem simpler than a straight line, whose ends are arbitrary. Touch is also a short-range sense. It cannot inspire feelings for spatial infinity. Indeed, most Greek symbolic spaces remained finite—at least in the classical Greek imagination.*

A tremendous enlargement of symbolic space took place, however, in the Hellenistic era, an enlargement which shows that by this time the Greeks were able to overcome the restrictions of tactile space—if such restrictions were indeed important earlier. For it was in this era that the geometry of Euclid was extended from abstract logic to visually observable cosmic distances. This astonishingly bold mental leap, the first application of geometry to the real three-dimensional cosmos, was made by Aristarchus of Samos (about 310–230 B.C.). Aristarchus was the first who attempted to determine the relative sizes of the earth, moon, and sun with the help of observations and using the laws of geometry. I shall not go into the technical details of his measurements. These can be found in many texts on the history of science. He had no means of measuring angles exactly and, therefore, while his method was correct, his results were not very accurate. Nevertheless he was still able to prove that, while the moon was smaller than the earth, it was much larger than anybody had ever sup-

* Not even the dwelling space of their gods was imagined to be in some faraway nowhere, like those of many other religions; rather it was on Mount Olympus, well within walking distance so to speak.

posed. He also discovered that the sun, contrary to visual appearance, was much larger than the earth, and thus he was the first to establish the cosmological insignificance of the earth.

And, as if these firsts were not enough, Aristarchus was also the first who assumed clearly and unequivocally that the sun was at rest in the universe and that the earth circled the sun. More than that, he also assumed that the reason why the motion of the earth does not produce the illusion of a movement of the fixed stars is that the latter are so much farther from the sun than the earth. This assumption was so bold and alien, the spatial imagination so daring, that not only did the contemporaries of Aristarchus refuse to believe it but, seventeen hundred years later, Tycho Brahe, the most diligent and successful observational astronomer of the sixteenth century, rejected the then emerging Copernican theory because he could not believe that such distances could exist.

In the works of Aristarchus and two of his outstanding successors, Eratosthenes of Cyrene (275–194 B.C.) and Hipparchus of Nicaea (about 162–126 B.C.), Greek astronomy far surpassed anything ever achieved before. One of the most exciting of all symbolic spaces, the large-scale, three-dimensional cosmic space with immense yet measurable distances, sizes, and proportions, was actually born in the imagination of these astronomers when they imposed a mathematical structure on the cosmos by projecting the rules of earthborn geometry far beyond the earth.

The same combination of theory and praxis, the same ingenious mixing of deductive methods with observation which emerged in Hellenistic astronomy, also characterized the best of Hellenistic science, mathematics, and art. One of the greatest intellects who ever lived, Archimedes, was active in this period and made astonishingly advanced and original contributions not only to deductive geometry and mathematics, but also to experimental physics and engineering. The translation of some of his works into Latin in the twelfth century was one of the decisive events in the evolution of rational scientific thinking in Europe. In its final period, Greek science erected a standard which was not reached for another fifteen hundred years.

The trend away from imposing preconceived perfection on the real world and toward observation and realism is also reflected in the visual arts of the Hellenistic era. Like Archimedes or

Aristarchus, the artists of the Hellenistic era were more concerned with observation and with a searching for realistic expression than their Classical Greek counterparts. They attempted to capture realistically the dynamics and tensions which make the world—while not perfect—at least interesting. A characteristic and famous example of the art of this era is the *Nike of Samothrace* or *Winged Victory* which now stands at the top of one of the main staircases of the Louvre (Fig. 4-5). It depicts an imaginary creature, yet is more alive and realistic in appearance than the earlier Aphrodite. It is able to communicate, in fact it radiates, the exuberance and the joy of victory. Its magnificent wings and masterfully sculpted drapery depict a sweeping motion. More important, it conveys an internal stress which was nonexistent in the rationally constructed harmony of the Classical era. Such original works, again, were not to be seen for another millennium and a half.

The decline of Greek civilization was gradual. Part of Greek culture was incorporated into Roman society. The Romans themselves, however, contributed relatively little to science, art, or philosophy. They were doers, not thinkers; they preferred action to contemplation, engineering to astronomy, politics to philosophy. Their accomplishments in science and in the arts, however respectable in themselves, seem to be, from our historical perspective, just the afterglow of the Greek radiance. Then, as the Roman empire declined, the intellectual level of Europe sank further, and from around A.D. 600, not only was the Greek artistic and intellectual spirit dead, but most vestiges of the Greek accomplishments had disappeared from European consciousness.

There were some important exceptions however. Among them was the cosmology of Ptolemy, which survived unchallenged as the accepted view of the solar system in both the European and Islamic civilizations throughout the Middle Ages. Claudius Ptolemy lived in the second century B.C. and was the most influential astronomer of the Hellenistic period. He worked out a detailed mathematical description of the movements of the then known bodies of the solar system based on the assumption that the center of this system was the earth at rest. Needless to say that while it is quite possible to describe the motion of the sun and the planets

Figure 4-5

based on this assumption, the resulting orbits of these bodies are rather complicated. Ptolemy's system therefore gave useful prescriptions to astronomers, but since the calculations were long and very complex, its application to the practical problems that arose in navigation and calendar making was limited. Aristarchus' sun-centered model, in contrast, was much simpler, but it lacked detail, and consequently, there was no way to put it to practical use.

With the disappearance of most of the Greek ideas, the cosmology of the early Middle Ages became, in many respects, similar to those of earlier, non-Greek civilizations. The earth was viewed once again as flat and as the center of the universe. Space was thought to be finite; it was believed to be enclosed in crystal spheres and to be in direct contact with the netherworlds of heaven and hell. The homogeneity and rationality of Greek geometrical space gave way to other symbolic spaces which were organized by religious symbols and values, and divided into sacred and profane places and regions. Maps were drawn fairly often, but even when the terrain to be represented was well known, they did not depict spatial relations faithfully. Allegorical symbols and relationships were far more important than realistic depictions of distances and directions.*

In the early Middle Ages, the techniques of both painting and sculpture were primitive by Greek standards. Painters were not interested in observing, and even less in depicting, the real world, and they were, either by choice or circumstance, ignorant of Classical art. The extension of space, for example, was usually ignored or substituted for by a flat golden background. Individual figures did not relate to each other in space; each existed separately in a continuum without scale. As in the art of ancient Egypt, there was little logic and less realism in these paintings. No one sculpted freestanding human figures which depicted motion. No one studied the forms and proportions of the naked human body. Medieval artists, like their predecessors in Egypt or in Babylonia, were quite capable of creating beautiful and expressive works. But they did not aim at the faithful reproduction of the visible.

* A hilarious yet not altogether exaggerated example of early medieval spatial thinking is described by Mark Twain in *A Connecticut Yankee in King Arthur's Court* in the scene where Sandy describes the whereabouts of a half-imaginary, half-real castle.

This they may have considered irrelevant. Instead, they attempted to convey the religious significance of the real and the imagined. Their figures were meant to exist not in the three-dimensional space of the senses, but in the symbolic spaces of the legends of Christianity. In medieval art, therefore, as in that of most ancient civilizations, there was little distinction between the reality of perceptual space and that of the symbolic spaces of the imagination.

Perhaps the only discipline in which the reality of the three-dimensional space of the environment had to be faced was architecture. Buildings had to stand in the real world, after all. And since many of the Classical buildings were still standing, their example could not be ignored. And where Classical techniques and spatial forms were first used to express Christian thinking through Christian symbolism, we see the emergence of the first characteristically Western constructions, the first masterpieces of European civilization: the Romanesque churches and monasteries.

The medieval period also saw a regression to a more primitive concept of time. Cyclical time came to dominate life and the imagination. Events were again thought to repeat themselves on a regular basis. Mass hysteria, for example, accompanied the expectation of the Second Coming as the year A.D. 1000 approached. In this atmosphere, astrology and magic flourished to an extent which would have dismayed the Babylonians.* Temporal relations were perceived in the same illogical way as spatial ones. Medieval painters matter of factly depicted biblical events not only as taking place in their own cities, but during their own eras. Chroniclers were as hazy about temporal as they were about spatial order. "When a medieval chronicler mentions the King," to quote an example from Lewis Mumford, "it is sometimes difficult to find out whether he is talking about Caesar or Alexander the Great or his own monarch: each is equally near to him." There was little sense of history as we know it, few attempts to place events in well-defined regions of space and time or to compare and relate to each other events taking place at different times and in different places.

* Sample: An important practical task of medieval astrologers was to find ways to determine, using her horoscope, whether a woman's "honor was corrupted."

5

Law and Order in the Flow of Time

(Polyphonic Music and the Scientific Revolution)

. . . music is given to us for the sole purpose of establishing an order in things including particularly the co-ordination between man and time . . .

Igor Stravinsky

It was only as medieval Europe gradually became acquainted with Greek thinking that slow but far-reaching changes started to take place in the prevailing European view of the world. That Greek thinking survived at all and was disseminated was largely due to the Islamic civilization which flourished from about A.D. 800 to 1200. The Arabs and the Persians discovered and appreciated Greek philosophy, mathematics, and science; rescued and translated manuscripts and commented on them; added their own important new discoveries; and generally maintained an atmosphere of intellectual curiosity and an interest in the real world.*

* Those accomplishments of Greek civilization which did not interest the Arabs, such as Greek art (the making of human forms was forbidden in Islam), were probably lost for good. Most of what we now know about Greek art comes from the often less than perfect Roman copies unearthed in Italy from the fifteenth century onwards.

Through contact with the Islamic civilization, the natives of Europe, beginning in the twelfth century, became acquainted with the most important Greek works, those by Aristotle, Euclid, and Archimedes. During the next three or four centuries, they assimilated Greek learning, and this assimilation led to the gradual reappearance of rational thinking.

The first results of this change made their appearance, not surprisingly, in fields that the Greeks themselves were most successful at exploring, namely in mathematics and geometry, statics and astronomy.* And the first great successes in science occurred in the area of the Greeks' greatest success: descriptive astronomy. The first great step—and the single most important step—in astronomy was the rediscovery of Aristarchus' model of the solar system. In the middle of the sixteenth century, Nicolaus Copernicus reinvented and vastly improved Aristarchus' sun-centered model. Copernicus turned Aristarchus' bold idea into a coherent observational-geometrical system supported by detailed calculations. About half a century later, Johannes Kepler improved upon the Copernican system with his discovery of the descriptive geometric laws governing the motion of the planets. Kepler's first law, for example, stated that the planets' orbits have elliptical shapes with the sun located at one focus of the ellipse. In the works of Copernicus and Kepler, Europe finally surpassed Hellenistic—and all other—astronomy both in accuracy and in consistency. The space of the sky and the motion of the solar system became simple and orderly, and gone were the unpleasant complications of the Ptolemaic system.

The concept of a sun-centered solar system changed the framework of human imagination forever. The large-scale symbolic space of the universe was hereby reordered; the earth and humanity were no longer at its center. As this new idea began to spread, it led to a revolutionary transformation of the entire human cosmology.†

* The study of mathematics was aided immensely when—again with the Arabs as intermediaries—the Indian positional system of notation was introduced to Europe.

† Indeed the very word "revolution" (which actually means "rotating around") came to signify an upsetting of the established order because the Latin title of Copernicus' work describing the revolving skies was *De Revolutionibus Orbius Coelestium.*

Yet with all their revolutionary implications, the works of Copernicus and Kepler, taken by themselves, still remained very much within the Greek modes of thought. The new picture of the solar system still contained nothing but spatial shapes and geometrical laws, and was still dominated by spatial concepts.* It contained no clue to the solution of the one outstanding problem that had forever eluded the Greeks, the Arabs, and the thinkers of all other great civilizations: a consistent mathematical description of temporal change, the change of spatial position—that is, of the motion of bodies.

This was an enormously important problem, and when its solution was finally found, human history took a new turn. As the historian Herbert Butterfield put it: "Of all the intellectual hurdles which the human mind has confronted and has overcome in the last fifteen hundred years, the one which seems to have been the most amazing in character and the most stupendous in the scope of its consequences is the one relating to the problem of motion. . . ."

The individual who first understood how motion could be described mathematically was Galileo Galilei. Galileo's popular fame rests chiefly on his being persecuted for his advocacy of the Copernican system, on his first use of a telescope to discover that there are bodies in the universe not visible to the naked eye, and on his invention of the pendulum clock. But if Galileo had never accomplished any of these things and had never spoken a word about the motion of the earth, he would still be considered the founder of modern science because his discovery of the law of the free-fall of bodies made him the first person to describe motion mathematically as a process in time and space and the first to check experimentally the validity of that description.

The most important conceptual breakthrough in Galileo's ap-

* It is not difficult to imagine that, given another peaceful century or two, Hellenistic astronomy would have discovered most details of the Copernican system. Such a discovery, however, would not have had the revolutionary significance of the Copernican idea. The ruling religions of the Hellenistic era had no strong beliefs concerning the structure of the skies. What gave the Copernican discovery such a tremendous importance was that it clashed head-on with the prevailing Judeo-Christian cosmology, which placed the earth and the human race at the center of the universe. The mental energy generated by this clash fueled to a large extent the evolution of modern civilization.

proach was the realization that all the important features of motion—the distance covered, the speed, the change of speed—could be expressed in terms of the time elapsed. To put it more succinctly, Galileo saw that *time is the independent variable in the description of motion.* This was an extremely significant and fruitful idea. First of all, it enabled Galileo to experimentally verify his law of the free-fall with relative ease. But it also implied an entirely new conceptual view of the world. In this view, the passage of time was a sovereign and fundamental process of nature, not conditional on anything else in the environment. This implied that motion had to be described in terms of time, not time in terms of motion. It was also implicit in Galileo's approach that the flow of time was uniform, i.e., could be mathematically regulated. Time could not have been the independent variable otherwise.

Eighty years after Galileo's discovery, Isaac Newton codified this concept of time for physics: "Absolute, true and mathematical time, of itself, and from its own nature, flows equably without relation to anything external. . . ." Newton's definition implied both that the passage of time was independent of the environment and that its flow was measurable.

To us today, these ideas seem self-evident. We read, for example, that the earth makes a full revolution around the sun in about 31,556,925.9747 seconds, or that the elementary particle called the "neutral pion" has an average life expectancy of 89 times the billionth of a billionth of a second. We marvel at the technical expertise involved in such accurate measurements, but we take it for granted that long or short time intervals can actually be measured. We also take for granted the assumption that these measurements will always remain the same: that they will be the same during the long, warm days of summer, during the short and cold days of winter, whether we are young or old, happy or unhappy, excited or calm.

The reason we take all this for granted is because we live in a society where social life is structured and regulated by accurate clocks—where measured time pervades most aspects of life. In our society, we have adjusted our life to a particular symbolic time we call "metric time" which is determined, structured, and measured by the numbers on our clocks. In industrial society, metric time dominates all thinking about time so strongly that when, for example, our own subjective "experience" of time, the "time we feel," contradicts metric time, we call the time of our own sensations an *illusion* and the number-based symbolic time *real.* Thus, to turn to a well-worn example, we feel that we have spent an eternity in the dentist's chair and merely a few minutes with our lover. But when we look at our watch and it tells us that both events lasted a half hour, we don't hesitate to describe the symbolic time of the clock as real and our real time perception as illusory.*

Metric time is also independent of the solar system. We don't set our watches by the sunrise; rather we say that sunrise occurred at such and such time. By divorcing time measurement from environmental events, we have been able to create such diverse symbolic notions of time as the time elapsed between the birth and death of the sun or the metronome-indicated time on Beethoven's Piano Sonata, Op. 110.

This concept of a measurable, independent flow of time seems so natural to us that it takes an effort to remember that it is a fairly recent invention of Western civilization. A thousand years ago the metric symbols of time did not exist at all for shorter time intervals and were rather vague for longer ones. And it is not difficult to understand the reason why. No matter how simple and self-evident metric time seems to us today, it is actually a complex and totally abstract mental construction removed from and even contrary to every internal and external human experience. What we take for granted now was a revolutionary notion that first made its appearance in the thirteenth century and became one of the cor-

* We display similar reactions when our visual estimation of lengths contradicts measurements. If, as in the famous Muller-Lyell figure,

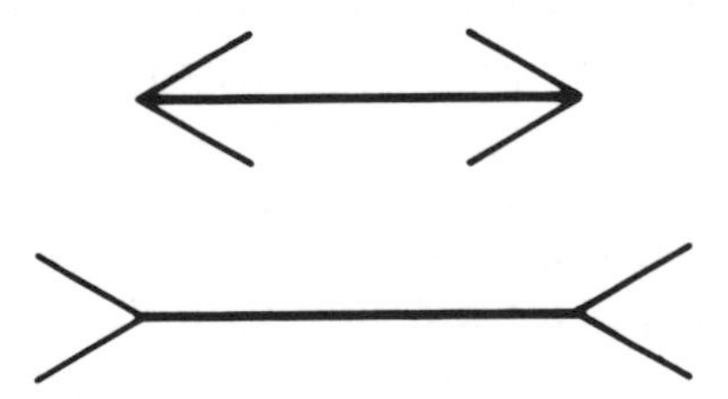

measurement shows that the two horizontal lines are of equal length in contradiction to what we see, then we call our experience an optical illusion. In societies less confident of the significance of numbers, such an experience would perhaps have prompted essays about the strange behavior of measuring rods.

nerstones of Western civilization. For without metric time, science, technology, and industry could never have come into existence.

It is fascinating to speculate on the origin of this powerful symbolic notion of time. How did it evolve? What factors prompted its appearance? Why did it appear in Europe only—and at a time when Europe was still quite backward compared to many other civilizations?

Such questions seldom allow for simple, clear-cut answers. But, as we will see later, one important and most likely decisive factor in the evolution of the idea of metric time was the extraordinary development of Western music. It was in the theory and practice of a uniquely Western musical form, polyphonic music and its measured notations, that metric time was invented, studied, and utilized for the first time in history.

I shall return to the role of music a little later. First, however, let us take a more detailed look at why the seemingly simple idea of exact time measurement is such an abstract and difficult notion. A good way to appreciate this is to compare the measurement of "time" with that of "space" and note how different these operations were, both in evolutionary and in cultural history. It turns out that the exact knowledge of spatial measurements was very important in biological and in sociocultural evolution, but the precise knowledge of the measures of time was not.

Strictly speaking, in the prehuman stages of evolution there were no "measurements" of any kind.* But even in animal life there had to be estimates or guesses of distances. Many higher animals jump to safety or to catch prey, and cover a variety of distances. An animal cannot survive without having either genetic or learned biological programs for fairly good and accurate assessments of distances. These assessments have to be quite accurate because they determine the motoric responses and the actual movement of the body. Distance guessing must have been particularly important in the human evolutionary branches. Our arbo-

* By measurement I mean the determination of the magnitude of a quantity by comparing it with standards or units of that quantity. It is important to realize that there cannot be measurements without units or standards, or without comparisons. Units and standards are, of course, human symbols, and "measurements" did not exist before human beings created them.

real ancestors could not have survived without being very good at guessing the distances between the limbs of trees.

Genetic programs also exist for other, more complex spatial measurements. The biological program known as the "territorial instinct" allows a creature to assess and recognize segments of space as its own. This requires a recognition and assessment not only of distances, but also of surfaces and, as in the case of birds and perhaps fish, of volumes. The pervasiveness of this sophisticated process in the animal world indicates how important the recognition of spatial extensions must have been in the later history of evolution.

Along with the ability to judge spatial extensions, the ability to estimate speeds of moving bodies (a menacing foe or a fleeing prey) was also supremely important in adaptation. Since the perception of different speeds triggered different motoric responses, the estimation of speeds also had to be quite accurate.

In contrast, there was no biological necessity for actively estimating or guessing the lengths of time intervals, and no life-form other than homo sapiens is known to do it. Time is organized in life by genetically programmed biological clocks that mirror environmental events and prod organisms into action at certain specific times. The biological clocks do not assess or estimate durations—that is not their function. They turn the paramecium on, for example, and make it mate at the same time of day every day. They govern our sleep and turn us on—if that is a fitting expression—to go to sleep. There is no active estimating or guessing of the lengths of the time intervals involved in these processes at all.

In Jerison's theory of the evolution of the mammalian brain (as we can recall), the first enlargement of the brain came about as the result of a need to translate temporally extended information into a spatial map of speeds and distances. Temporal measures of durations or frequencies were interpreted in spatial terms because the spatial patterns had much more significance for mammals than the purely temporal ones.

These evolutionary developments were paralleled in human cultural history. Measures of space evolved early on because their use was a social necessity: neither agriculture nor commerce could develop, and the artifacts of hunting and war could not be made, without reliable and fairly accurate methods for measuring lengths, surfaces, and volumes. But there was no real social necessity for accurate measurements of time, and of short durations in particular, until relatively recently.

What human societies needed from early on was an ability to *keep track of time.* This is often confused with time measurement although the two operations have nothing in common. Keeping track of time simply means adapting to the phases of a periodically changing environment. To help in this process, calendars and clocks of varying complexity were devised in all civilizations. But these clocks and calendars were used by societies in much the same manner as biological clocks are used by living creatures. Calendars helped societies adapt to such periodic environmental events as the changing of the seasons, and clocks indicated the time to go to work. They mirrored the environment and allowed society to prepare for certain foreseeable future events. They did not in any way measure time.

Nor did clocks or calendars aid in the establishment of the idea of the uniform, regular flow of time. On the contrary. The insoluble problems inherent in the construction of an exact calendar may well have inhibited the perception of time as having any rational or reliable measures. We often say that the solar system behaves like clockwork and to some extent we are right. But the solar system is also a mad clock beyond repair. None of its periods can ever be made to "fit" accurately into any other. There is no "year" which contains an integer number of days or an integer number of months, nor is there a "month" with an integer number of days. The "day," which is, after all, one of the dominant time intervals in human life, cannot be used as a permanent time standard, except at the equator, since its length keeps changing throughout the year. Since the periodic mechanisms of the solar system were relied upon as the main agents in defining the passage of time, societies could hardly have evolved an intuition for metric time. The Pythagoreans despaired when they could find no number which measured the length of the diagonal of the unit square. Calendar makers encountered such problems throughout history, but their troubles were taken for granted.*

* The idea that the mechanics of the solar system were such that an exact calendar was impossible in principle was proposed repeatedly. The best-known speculations in this direction are those of Nicole Oresme, the most important mathematician of the fourteenth century. Ironically, he was also the author of the metaphor of the "heavenly clockwork."

The physiology of our own bodies does not help us to grasp the idea of rational and permanent time standards either. The life process offers two basic standards for the judgment of short durations. One is the intake and expiration of the breath, the other is the heartbeat. Neither is reliable. Both are influenced by differences in age and emotional or physical state. Our physiology does not offer reliable time standards for longer intervals either. Individuals evolve, mature, and age differently and have different life spans.

In fact, all possible *perceivable* time standards, whether found in our external or in our internal environment, are nonspecific, constantly changing, without any common measures at all. Neither nature nor nurture, in other words, would have evolved the intuition that the flow of time could be regulated by exact, objective, reliable standards. To appreciate the significance of the lack of such perceivable standards, one only has to imagine a world where all external objects, as well as our own bodies, arms, legs, keep changing their lengths. Some change regularly, others haphazardly, and only some loose, general laws regulate these changes. It is difficult to imagine that in such a world exact standards of length or distance could ever have acquired a meaning. Mathematicians study such imaginary worlds in topology. These mental worlds have interesting properties, but ideas like length or volume simply do not exist in them. These are worlds without metric, and the world of human individual and social time before the "invention" of metric time was much like them.

It is sometimes assumed that metric time became established with the invention of mechanical clocks. This is, however, not so. The first such clocks were constructed in China around the ninth century. Yet the idea of metric time arrived in China from the West, together with industrial civilization, almost a thousand years later. In Europe, the first mechanical clocks appeared in the early fourteenth century. But, as in China, none of these early clocks were used to "measure" time in the true sense of the word. These clocks were, first of all, not accurate enough to reliably define standards for short durations. Their accuracy was not significantly better than that of the processes already available in the environment. Actually, "showing the time" was only a secondary function of these clocks. One of the first elaborate mechanical clocks in Europe was built by Giovanni de Dondi in Padua in

1364. "It contains seven dials showing each of the planets, and all sorts of other astronomical data with an extra rather inconspicuous dial that tells the time," writes the science historian Derek J. de Solla Price. The symbolic function of modeling the solar system was, in other words, the all-important role of the early clocks, and it was their ability to do this which generated such excitement at first. In no sense were these clocks used for genuine time measurement. Price, incidentally, saw this clearly, and he wrote about the necessity to "disentangle the clock from the history of time measurement and connect it instead with the longer and earlier history of astronomical models. . . ." Clocks could measure time only *after* the idea of a measurable time became established. By modeling the motion of the sun, the early clocks did little more to promote the notion and the perception of metric time—either in China or in Europe—than did the observed motion of the sun across the sky itself!

The early clocks, however, whether mechanical devices or hourglasses or, as in the Far East, candles, did play a role in regulating the rhythm of social life and in detaching it from the rhythms of the natural environment.* This type of timekeeping evolved to a high degree of perfection in medieval monasteries, where it was used to regulate religious observances. Some historians see in the process of the detachment of social rhythms from environmental rhythms the source of the temporal regularity which characterizes modern industrial civilization. This may well be so, but again, like biological clocks, the early time markers merely regulated behavior. They did not use units of time, nor did they provide for the comparison of different durations. They never actually measured time.

When Newton described absolute time as something which flows "without relation to anything external," he emphasized the independence of the flow of time from the environment. It seems that this view of time—like the metric—must also be learned. When Jean Piaget investigated the evolution of time perception in young children, he found that even in our industrial society it

* In the Far East, special candles were used for regulating behavior. These were marked off into equal segments, and each segment was labeled with an appropriate symbol to indicate that the time for this or that activity had arrived.

requires considerable maturity for a child to be able to disentangle the perception of the flow of time from the environment and from the more fundamental perceptions of speed and distance. The latter two dominate the perception of all young children at first. At an early age, reports Piaget, "All temporal judgements are . . . actually disguised spatial judgments. The [temporal] order of events is confused with the [spatial] order of points of the paths, duration with space covered and so forth." Piaget also concluded from these experiments that, at first, the perception of time is always derived from the perception of motion. The passage of time is perceived not in the abstract but through the motion of bodies. If, for example, an experimenter moves different bodies at different speeds, young children cannot say which body moved for the longest time or whether they stopped simultaneously. "We merely have to make all speeds unequal for all temporal intuition to be falsified," wrote Piaget. Only at a later age can children conceive of the abstract passing of time independent of events in the environment.

The history of conceptual thinking about space and time shows an outline which is remarkably similar to the evolution of biological, social, and individual perceptions. The first important rigorous laws to be discovered and applied—those of Greek geometry—had to do with spatial properties. When the first known philosophers speculated on the nature of time, they attempted to understand the flow of time by connecting it with the motion of bodies—just as young children do. Thus Plato, as we recall, tied the very existence of time to the motion of the heavenly bodies, and Aristotle attached it more generally to the motion of any spatial object. Time, for Aristoteles, could not be separated from movement.

It seems that the only important thinker in the entire prescientific era who thought of time as being both independent of motion and measurable by itself was St. Augustine, the fifth-century bishop of the North African city of Hippo who was one of the most original philosophers of time ever. He was, among other accomplishments, the first to recognize that the idea of cyclical time was incompatible with the spirit of the Bible. He rejected explicitly Aristotle's notions of ever-recurring cycles and declared that not only the creation of the world but also the salvation of humanity by Christ was a unique, never repeatable event in history. He advocated correspondingly that history was open-ended, and ridiculed astrology.

St. Augustine's time concept also included the idea that time cannot be older than the universe—i.e., that time started when the universe started—a very modern notion indeed. What is most important, however, is that all the essential notions of independent and metric time can be found in St. Augustine's writings. He advocated clearly and concisely the concept of a "time" which was autonomous, not derived from the movement of bodies, whether heavenly or earthly—a concept very similar to Newton's "absolute" time. Augustine also had the idea that time was measurable and he proposed an ingenious way of performing pure time measurements by reciting classical, quantitatively defined poetic meters. He seems to have realized that only through hearing can we perceive temporal regularities without confusing them with spatial attributes. For Augustine, time was, therefore, independent of space and motion and was also metric.

In spite of his great theological authority, virtually all of St. Augustine's ideas on time were ignored throughout the Middle Ages. Medieval scholars were much more influenced by the Aristotelian ideas about time and therefore viewed it as a less than fundamental dimension whose existence was derived somehow from motion. The rejection of St. Augustine's ideas throughout the Middle Ages is nothing if not a convincing illustration of how ineffective cognitive thinking was, particularly in earlier times, when it challenged biologically and socially determined intuitions.

Despite all these obstacles, despite all the perceptual and conceptual difficulties and the evidence to the contrary presented by biology and society, metric and independent time was to become the dominant symbolic time of Western civilization. In fact, there is reason to believe that by the time Galileo solved the problem of free-fall both experimentally and theoretically by introducing time as an independent and measurable quantity, this notion of time was already well established among educated Europeans. The best evidence for this belief is that Galileo's revolutionary new treatment of time was accepted without a murmur by his quarrelsome contemporaries, many of whom were opposing just about everything else he said. Furthermore, even Galileo himself did not make much of his discovery. He treated it as just another clever mathematical trick and never seemed to realize its revolutionary character.

The solution to this curious historical riddle might be this: that metric and independent time *was* a common intuition by Galileo's time and that this was true because, in the four centuries preceding Galileo, an important sensory model had been created in Western Europe by which a numerically well-defined, exactly measurable, and exactly measured structure was imposed upon the experience of what we call the flow of time. This model had nothing to do with science or philosophy. It was the theory and praxis of polyphonic music and its measured notations. Their effect on the general intuition of time must have been considerable since virtually every educated person in Europe was familiar with them. The study of music theory, like the study of geometry or astronomy, was compulsory in higher education. A short survey of the history and nature of polyphonic music will be all we need to establish the extraordinarily important part this musical form played in the evolution of our civilization's symbolic time.

We call a piece of music *polyphonic* if it combines two or more simultaneously sounding melodic lines. In one sense, polyphony is as old as human singing itself. In group singing, for example, the voices of men are always separated from the voices of women and children by at least one octave. Singing in "parallel" fifths was also practiced in various cultures, as were perhaps canon singing and similar forms of simple polyphony. But here I shall use the word "polyphony" to describe music in which genuinely different melodies are sung or played simultaneously according to a consciously organized system. This kind of music arises from the composer's desire to explore ways to create new music by combining and intertwining two or more melodic lines. This type of music evolved in Europe in a slow but steady manner from the eleventh century on.

Only the rough outline of this evolution is known. The musical raw material of early polyphony was a collection of melodies known as the Gregorian chants.* These austere and subtly beauti-

* They were named after Pope Gregory, under whose sponsorship they were first collected and catalogued.

ful melodies formed the musical background to the services of the Roman Catholic Church throughout the Middle Ages, and the manner in which they were sung was prescribed by religious law. The words of these chants were Latin translations of the Psalms and originated in Jewish liturgy. Their melodies had more diverse origins, although the majority were also from the psalmody of Jewish religious services.

At first, the melodies were sung either monophonically, with everyone singing the same melody at the same pitch, or in parallel octaves. Sometimes they were also sung in other parallel intervals, usually fourths or fifths. This type of singing was prevalent in the early Middle Ages and became known as "parallel organum" or "strict organum." In such a piece we hear no new melodies but an enriched sonority.*

The next stage was the evolution of a more sophisticated form of organum which is sometimes called the "free organum." This innovation emerged during the second half of the eleventh century in France and in England. In the new organum, the rules of the parallel motion of melody lines were gradually relaxed, and variable intervals between the parts were introduced. The distance between the pitch of two melodies, which might start out as, for example, a fifth, could then change at some point to a fourth or to a third. As a result, the different voice parts no longer moved only in parallel. They went through various separations and sometimes even moved in opposite directions. One melody went up while the other came down. But there was always an original melody at this stage—the already known melody of a Gregorian chant. This basic melody was called the "cantus firmus" or fixed melody. In the free organum, this was sung sometimes above, sometimes below the other melody. The voice parts, in other words, sometimes diverged, sometimes converged, and they could cross each other.†

* The most important document about the musical practices of this era is a collection of musical examples from the ninth century called *Musica Enrichiriadis.* It consists mainly of parallel organa together with instructions as to how to sing them.

† The largest source of information about the eleventh-century organa is a collection of manuscripts at Corpus Christi College at Cambridge University. This collection is known as the Winchester Tropes and contains more than 150 two-voice compositions.

Further innovations started to emerge in the twelfth century. Their first records of these developments were found in the monastery of St. Martial at Limoges in central France. It seems that a new style of singing, which later became known as "melismatic," or "florid," organum evolved there. In the earlier forms of the organum, the two voice parts always provided a single note against a single note. Even when the melodies were genuinely different, the durations of the simultaneously sounding notes were the same in both parts. The new development consisted in a relaxation of this rule. It became possible, in the melismatic organum, to sing the second melody, the one which was added to the cantus firmus, in a freer way. In this style, a whole group of notes could be sung in the second melody against a single note in the original chant. The lengths of such a group of notes varied from a few notes to long variations, or melismas, containing ten or twenty notes. The singers of the cantus firmus then had to sustain their notes until the singers of the variations were through with their independent parts. This was already a style very different from either the parallel or the free organum.*

In the late twelfth century, a most important school of polyphonic music came into existence in Paris. It evolved at the Cathedral of Notre Dame in Paris, which was then in the process of construction (the actual construction began around 1165). The members of this school were the musicians of the cathedral, but it is likely that the Notre Dame school also included an international group of active music theorists from the University of Paris. The rise of the school was simultaneous with the rise of Paris itself, which was soon to become the leading cultural and intellectual center of Europe. These avant garde musicians of the city, its main church, and its great university played a great role in estab-

* History, as usual, has done everything possible to confuse the meanings of the various terms. Thus it was the florid or melismatic style which became known under the name "organum," while the older style, in which the principle of note against note still more or less prevailed, became known as the "discant" or "discantus" style. The discant style remained much in use, and from it came a synonym for polyphony, the word "counterpoint." It came from the Latin expression for "note against note": *punctus contra punctum.* In modern usage, however, just to confuse things further, the meanings changed again, and the term polyphony is used today to describe early medieval music, while counterpoint refers mainly to later sixteenth- and seventeenth-century polyphonic music.

lishing this cultural leadership. Two successive organists of Notre Dame during this time were the first two composers in history whose names were associated with their individual compositions! They were Leonin and his successor Perotin. The latter was so admired that he was called "Perotin le Grand."*

What is known about these two composers and their works comes from a theoretical treatise which dates from the late thirteenth century, i.e., from a full century later. The writer, whose name has not been recorded and who is known to historians as Anonymous IV, was probably a student at the University of Paris. His treatise mentions the names and the works of Leonin and Perotin, and remarks that Perotin "was a greater composer than Leonin." Perotin was probably the first composer who wrote polyphonic works with three and four simultaneous voices. This was a revolutionary step which, in itself, would make him one of the most original innovators in the whole history of music.

To sing such complicated works was an unprecedented challenge, and not everybody was able to cope with it. For a long time, musical practice was to compose choral works with both monophonic and polyphonic parts. The latter were sung by specially trained solo singers (who were rewarded with a double salary for their abilities), while the choir sung the easier, monophonic parts only. It was two centuries more before less-trained ears could cope with polyphony, and only then did choral polyphonic singing become general practice. This huge time span illustrates how difficult it was for the medieval mind to cope with the practice of more complex polyphony.

But the theory was not easy to comprehend either. Since the mere singing of these pieces was difficult enough, improvising on them in performance was impossible.† Such music had to be

* Nothing shows better how central a role music played in those days than the fact that the names and the works of these musicians have survived. In contrast, neither the names of the builders of the magnificent cathedral itself nor the identities of the sculptors of the exquisite statutes on its walls were recorded anywhere.

† Centuries later, the baroque masters of the organ (J. S. Bach, G. F. Handel, and D. Scarlatti among them) were able to improvise polyphonic pieces during a concert performance. Another couple of centuries saw jazz players improvising several simultaneous melodies and rhythms. But in the early days of polyphonic music, this was out of the question.

thought out in advance and written down in advance. And if composers wanted to hear their ideas, they had to find a way to write them down so that they could be performed correctly. Writing down the structure of polyphonic music was, however, the equivalent of finding a symbolic way of representing and comparing durations. The reason for this is the following: Polyphonic music consists of several simultaneously sounding voice parts and several simultaneously proceeding melodies. It is therefore a temporal structure whose components have to fit together. The composer of such a piece has, therefore, to think carefully about the temporal organization of each separate part as well as about the organization of the piece as a whole. But the composer cannot even begin to think about such problems until and unless the temporal profile of each melody can be regulated by the same time unit. Without inventing a time standard, the composer can never make the flowing melodies mesh properly. This does not mean that each of the musical notes must have the same duration. But all the durations must stand in a simple, clearly definable relationship to a basic time standard or to a unit time length. This is what allows a composer to compare different time intervals in various melodies or to divide such intervals into equal or unequal parts. Only this permits the preservation of order and coherence in the flowing temporal structure of the work.

Thus the composition of polyphonic music involved careful thinking about temporal units, their ratios and proportions. But mere thinking about time units was not quite enough. Polyphonic music was not a philosophical theory. A practical way had to be found to notate the temporal units and their various ratios. A practical, recognizable notational system for the time values of the individual voices had to be developed if polyphony was to flower in actual performances. The Notre Dame school created exactly such a system. Although it was very primitive by today's standards, the inventors of this system nevertheless took an historic step because they created the first symbolic comparison of simultaneously passing temporal intervals, and thus discovered chronometry: the rigorous addition, multiplication, and division of temporal intervals—*the symbolic manipulation of durations.*

The Notre Dame system of musical notation is now called the "rhythmic modes." The rhythmic modes are essentially predefined patterns of long and short notes. Like most new inventions, the system itself was neither wholly logical nor free of ambiguity. But it defined a standard of temporal duration (called *breve*) as well as its double (*longa imperfecta*) and its triple (*longa perfecta*).* Using these well-defined durations, the early music theorists then defined six different "modes"—i.e., prescribed patterns of the one short and two long notes. These were later named after the Greek poetic meters which they seemed to resemble. The first mode, for example, was called the "trochaic" mode. It was the most widely used and it consisted of units made up of one double-length note (*longa imperfecta*) paired with one short one (*breve*): long-short, long-short, long-short, and so on. Another mode was called dactylic, and it was a pattern of long-short-long, long-short-long, where the first long was always perfect and the last imperfect. An extra symbol indicated how many times a pattern was to be repeated.

Notations for rests were, of course, also included in this system. The specific durations of the rests were just as important in a rhythmic pattern as that of the sounds. Anonymous IV, whose treatises consist essentially of a description of the Notre Dame notational system, devotes a whole chapter to the notation of rests. He calls a rest "a pause or omission of sound for a definite length of time or durational unit. . . ." Sounds simple enough. But the symbols of rest are nothing if not instructions to measure time intervals independently of anything else. We know of no philosopher of the thirteenth century who could have written such a sentence. All were still mesmerized by the circular Aristotelian idea that bodily motion measures time and time measures motion.†

It is also worth remembering that musical chronometry was very exact. Not until the appearance of electronic instruments in the twentieth century was there a more sensitive device for measuring durations than the human ear.

* The Latin adjectives reflect the fact that, in medieval times, numbers were believed to have qualities. The number 3 was a more perfect number than the number 2 because all things must have a beginning, a middle, and an end, and also because the Holy Trinity was associated with the number 3.

† From a detailed study of medieval mechanics by M. Clagett of the University of Wisconsin, one can conclude that the main stumbling block which prevented medieval scholars from even approaching a solution to the problem of motion was their inability to consider time as an independent dimension of motion.

The invention of the notation of rhythmic modes was just the beginning of the evolution of the theory and praxis of true chronometry. Around 1260, Franco of Cologne, a leading music theorist of his time, published a book entitled *Ars Cantus Mensurabilis* in which the so-called *mensural* notations were defined. These new notations got rid of all the rhythmic modes at once. A piece of music could now be read without knowing the rhythmic modes in advance. With the mensural notations, chronometry—and music itself—became independent of poetry.* In this notational system, incidentally, there were seven fundamental time values with well-defined ratios: three longs, two shorts, and two even shorter durations (these were called *semibrevis*). In addition, Franco also improved the notation of rests.

Most of the early history of musical notation consisted of inventions of mnemonic aids which gave an approximate idea of the melody or of symbols which stood for recurring melodic fragments. After the need for adequate notation of music arose, the main stumbling block became always the notation of rhythms, or rather, of time intervals. The notation of pitch was a much easier problem thanks to our intuitive perception that some pitches are "higher" or "lower" than others. This intuition is quite universal, and it was not too difficult to find a way of depicting the degree of difference between pitches graphically. The use of a staff, or series of vertical lines, was already well known when the famous eleventh-century monk and music theorist, Guido d'Arezzo, improved upon it. Guido's system was, in principle, the same as the one we use today. In contrast, the search for a notation of durations was much longer precisely because of the many difficulties associated with the perception of measured time. But once time notations were discovered, their very existence profoundly changed music itself. New notations sometimes acquire a potential—mathematics is a good example—which allows them to extend their uses far beyond the problems which they were origi-

* Thus began a development which eventually allowed music to take autonomous forms in Western civilization. In traditional societies, music always accompanies some other activity: dance, poetry, theater, religious ritual, political ceremony. A piano sonata, a string quartet, or a symphony, in contrast, are meaningful in themselves.

nally supposed to solve. This turned out to be the case with musical time notations.

The most important new musical characteristic that evolved from the theory and praxis of time measurement was the musical meter. Meter gave the temporal flow of music a new structure. In rhythmically free music, like the Gregorian chant, "We have longer and shorter tones in temporal succession, and this succession is itself already the entire rhythm," writes the music theorist Victor Zuckerkandl. Since the discovery of notated music, we also "have longer and shorter tones in temporal succession; but here the succession also gives rise to the metrical wave, whose uniform pulsation is perceptible through all the changes. . . ." This uniform pulsation results from the presence of strict temporal units which discipline the free flow of rhythms. The easiest way to appreciate the difference between meter and rhythm is to think about reciting a verse. As long as we recite the line stressing the regular beats, the metric structure comes through. But the meaning of the verse gets lost. If we recite the same line stressing the meaning, we get the true rhythm of the verse. The metric constraint, however, emphasizes the verse's temporal structure; it makes the meaning more significant and beautiful.

The necessity for using exactly measured time in composition arose in polyphonic music only. Time units are often needed, in some sense, in monophonic music as well. A singing and dancing or a marching group, for example, cannot stay together without using the same time unit. They have to keep time, in other words. But, as I emphasized earlier, keeping time is not time measurement. Horses can be trained to keep time, but they cannot be trained to measure it. Measurement is a symbolic process; it involves a conscious comparison of simultaneously passing time intervals and the use of units or standards. It is precisely this operation that is involved in polyphony and in polyphony only.

The time structure of music continued to change; rhythmic and metric ideas were further developed by an important musical movement called the "Ars Nova" which came into existence first in France and soon afterwards in Italy. One of the characteristics of the new movement was its attempt to liberate musical time measures from the preconceived ideas of theologians and philosophers. The Ars Nova (which simply means "new art") got its name from one of a number of influential treatises written by the composer Philippe de Vitry (1291–1361) around 1325. Philippe, later the Bishop of Meaux, incidentally was a leading intellectual figure—one of the most cultured minds of the fourteenth century and a true Renaissance man before the advent of the Renaissance. A poet, mathematician, music theorist, and composer, he was a friend of Petrarch and was admired by and collaborated with such figures as the mathematician and astronomer Nicole Oresme; Jehan de Murs, an eminent mathematician-musician; and the Provençal mathematician and Talmudic scholar, also the inventor of mathematical induction, Rabbi Levi ben Gerson. All were busy with musical theory, which was by far the most exciting intellectual adventure of the time. A new world emerged in this evolving symbolism in which the temporal structure of musical motion was faithfully represented on a spatial axis. An immediate scientific "spin-off" of this preoccupation seems to have been Nicole Oresme's idea that one also could represent real motion graphically on a spatial axis. He used his idea to prove a simple kinematical theorem relating to accelerated motion.

But in spite of their great interest in theory, Ars Nova as a movement emphasized the importance of praxis. The new works are full of new ideas and new notations for rhythmic structure. They represent a fresh, bold, more intuitive, less conceptual view of music. As the practice of music was now being emphasized over theory, the composers were replacing the philosophically and theologically motivated theorists as the main exponents of musical ideas. As a consequence, even theoretical works began to be written with a more practical slant. Instead of speculating from "general principles" what was "correct" or "right" in music, the composers were more concerned with extending musical possibilities, with being able to write new-sounding, more-satisfying music.

This trend away from preconceived theory seems to be the first manifestation of a new attitude towards reality. When the musicians of Ars Nova decided that praxis was the test of the theory, they opened up a new path in Western civilization. A century later, the painters of the Renaissance discovered the same idea when they started to depict the world not as it "ought" to look, but as it was seen in reality. Another century and a half saw Galileo extending the same principle to the mathematical description of nature.

The new, practical approach led to further rhythmic refinements, which now were determined by the needs of the music and not by its philosophy. Thus the distinction between "perfect" and "imperfect" time values were abolished and, as a result, music got rid of the domination of triple time divisions and of the lower status of the "imperfect" intervals. Binary time divisions became fully acceptable. This single innovation alone greatly increased the composer's freedom in selecting the temporal structure of the work. As a result, music became more lively, more interesting, and more expressive. The composers of the era, the best known of whom were Guillaume de Machaut in France and Francesco Landini in Italy, were now able to produce and explore new and highly sophisticated temporal patterns. Machaut introduced "isorhythms," temporal patterns in which the same series of time values is repeated several times in succession, although the melodies may be different each time. The playful use of time-reversible melodies, melodies which also make sense if read backwards, became frequent. Fanciful rhythmic patterns like the *hocket* were created. In a hocket, the notes of a melody alternate rapidly with rests; notes and rests complement each other, giving rise to a sharp and complicated-sounding rhythmic pattern.

The creation and evolution of the style and the philosophy of Ars Nova aroused strong opposition. The Roman Catholic Church did not remain indifferent to the fact that much attention was paid to secular music and that even when writing devotional music many composers were more interested in the problems presented by the musical forms than in the way their music fit into the services. Early in the fourteenth century, Pope John XXI issued a papal bull condemning the practices of some musicians and objecting, in particular, to "some disciples of the new school" who "are greatly concerned with the measurement of time values."

But there was no turning back to the unmeasured time values and to the simple parallel organum. The problems associated with musical notations were far too challenging, the emerging new music was far too seductive. Instead of a turning back, the next two centuries saw the full development of vocal polyphonic music as well as the beginnings of sophisticated instrumental music. It was during this process that the musical center of Europe shifted from France to the Netherlands, where it remained until the seventeenth century.

Vocal polyphony came to its artistic peak during the fifteenth and sixteenth centuries. Large-scale works, both sacred and secular, were created in which several voice parts were woven together. The texts became completely secondary to the majestic beauty of the music, so much so that often the different voice parts were sung in different languages. The simultaneously sounding melodies—their composition governed by strict laws—evoked the spatial images of architecture. It sounded as if musical time spans had been fitted together to build monumental moving structures. Josquin des Pres in the fifteenth century, Orlando di Lasso, Pierluigi da Palestrina, and Thomas Byrd a century later, became the universally acknowledged masters of the musical arts.

The notational technique grew with the demands of music. By the end of the sixteenth century, an adequate theory of musical time measurements had developed. It allowed the denotation of durations in an abstract, symbolic way which was completely satisfactory in musical praxis. This system was, for all practical purposes, already identical to the one in present use. The present system is a simple binary one in which each type of (undotted) note represents a duration twice as long as the preceding shorter one. The dots are organized the same way. One dot after the note increases its duration by one half, the second dot adds another fourth to the original value of the note. All this seems to us simplicity itself. Yet it took about five hundred years of intensive thinking and trial and error experimentation before this system was found—an eloquent proof of how difficult it was to come to terms with the intuition of metric time.

Once acquired, however, the new intuition became quite common. The evolution of musical theory and practice from the twelfth to the sixteenth century was not an isolated enterprise which interested musicians only. The influence of musical ideas on the educated segments of the population during all this time was much greater than it is today. Both the theory and the practice of music was basic in medieval and Renaissance university education. A good part of the curriculum for all students who did not prepare for priesthood was a complex called the *quadrivium* which consisted of arithmetic, geometry, astronomy, and music. The four were thought to form a logical unity. Consequently, all students were exposed to the theory and the practice of music and of musical notation; all studied these highly sophisticated symbolic time measures and practiced to perceive them.After such studies, they needed philosophical treatises to help them perceive time as a measurable and independent variable about as much as they needed essays to learn that the air was breathable. We may well assume, therefore, that by the sixteenth century, the intuition of measurable time had become thoroughly common. And we need not be surprised that when Galileo pioneered the conceptual use of this notion of time, no one, not even himself, found the idea as revolutionary as we do from our longer perspective.

The motivation of the medieval musicians was neither scientific nor philosophical. They merely wanted to learn what kind of music could be created by combining several melodic lines and what laws govern these combinations. But this music made an impact far beyond the artistic sphere. Contemporary records do not reflect it since no one was conscious of it. The change was in perceptions, not concepts. Yet it was in the theory of polyphony that the problems of real time measurement were first studied and solutions applied, and it was in its praxis that an emerging civilization was educated to perceive the flow of time as a process which was not derived from the sun or the moon or from the motion of bodies or from any other primary cause, and which could be treated in much the same way as a spatial dimension. And since this perception was indispensable to the evolution of modern science, polyphonic music can claim to be one of the parents of modern science.

6

Classical Time and Space

(The Mammalian World with a Human Face)

. . . a world where beauty and logic, painting and analytic geometry, had become one.

Aldous Huxley,
After Many a Summer Dies the Swan

Our visual apparatus evolved to enable us to collect and evaluate information from faraway sources. But if we rely on the structure of our eyes alone, we cannot generally determine how far away those sources actually are. As a space-exploring sense, human vision is very good at perceiving forms, shapes, colors, and textures. It is very good at perceiving various degrees of illumination. It can also register motion and other changes extremely well. But where distance is concerned, the eye generally feeds the brain ambiguous and imprecise data. The reason for this lies in the structure of the eye. The eye operates like a camera in the sense that both contain lenses that focus an image of the external world on an inside surface. This surface is a light-sensitive film in a camera and the retina in the eye. After stimulating the receptor cells in the retina, light itself plays no further role in visual information processing. All the direct information that the incoming light provides about an object is contained in the image on the two-dimensional retinal surface. The image on the retina, however, only contains information regarding the *direction* and *apparent size* of the object. It contains no information regarding distance.

This deficiency in vision must have been a persistent problem throughout evolution.* To ameliorate it, adaptation invented two different genetic programs. The first is binocular vision, the presentation by the two eyes of two slightly different retinal images of an object to the brain, the degree of difference depending on the distance. The brain determines the distance of the light source by comparing the two images. This process is reliable, but only if the object is less than 15 meters away. If it is farther off, then our brains cannot distinguish between the two retinal images. In such cases, a second genetic program is called into play in which the brain scans the environment for "visual cues" to aid in its perception of distance. Such a visual cue would be, for example, the gradual diminishing of the apparent size of an object with distance. This, of course, is a cue only if the *real* size of the object is known. Other cues include the diminishing of the apparent distance between faraway objects and visual illusions such as the convergence of parallel lines at a point at a distance. There are many others.†

This genetic learning of the determination of distance based on visual cues was one of evolution's great achievements. What interests us here, however, is that distance judgment was one of those evolutionary problems which, after being solved genetically in adaptation, reappeared later in a symbolic form in human

* It still looms large on our visual horizon. Consider this problem in astronomy: In order to calculate how far a visible object is, be it a star or a galaxy, we need to know (1) how much light the object is emitting and (2) what fraction of this light reaches the earth. Since we can only measure the second, we can never know from appearance alone how far the visible object actually is. We must always find other data to substitute for the missing piece of information. If these data are ambiguous or uncertain, we are in trouble.

† A reader with some familiarity with astronomy will realize that the astronomer's methods of determining the distances of visible objects in outer space copy quite faithfully the biological solutions. For close stars, the parallax method is used. The principle is similar to that of binocular vision: the position of a nearby fixed star looks different from widely separated points on the earth's orbit. For more distant objects, visual cues such as the apparent brightness of the Cepheid variables (whose real brightness is supposed to be known) or the redshift are used.

cultural evolution. Aristarchus' determination of the size of the moon is a striking example of such a symbolic solution to a distance-judging problem involving large distances. Another symbolic solution to problems of distance perception emerged in the practice of the singularly human activity of representing the three dimensional world on a two-dimensional surface—i.e., in the practice of painting and drawing. This development had a profound effect on later human cultural evolution.

The peculiar property of paintings and drawings is that, although they are in fact two-dimensional, we can often interpret them as three-dimensional reality. The reason we can do this is that no matter what we look at, the retinal image presented to our brain is always two-dimensional. The retinal image of a painting, therefore, is essentially the same as the image of any three-dimensional object. And whenever the information reaching our eyes from the flat, painted surface is *essentially equivalent* to information we would normally expect from a three-dimensional view, we see three-dimensional space. In this case, our brains cannot tell the difference on the basis of vision alone.

Since the eyes of other developed mammals are not very different from the human eye, we would expect them to obtain much of the same optical information from a painting as the human eye. And yet no animal recognizes a painting or a drawing as a model of reality, no matter how naturalistic it might seem to us. The main reason for this deficiency is that animals cannot voluntarily disregard the sensory input informing them that the picture (or sculpture or movie for that matter) is merely an indifferent piece of cloth or marble. This intentional disregard of channels of information, the "willing suspension of disbelief," is a characteristic human ability necessary for the processing of such symbolic information as paintings, literature, or music.

Paintings (and artworks in general) are symbolic constructions in the same sense that novels or scientific theories are. All share the basic property of symbols: They are analogous with or refer to but are never identical with what they represent. All represent and re-create the world symbolically in the abstract. The space in a painting—no matter how naturalistic—is always an abstract, symbolic space. The space we see in the painting is not the same as the space we live in. We cannot walk, breathe the air, or measure distances in the space of a painting any more than we can in the space described in a novel or in the momentum-space of physics. While the latter are "made up" of concepts, words, and numbers, the space in a painting is made up of lines, contours, and patches of colors. To stretch the analogy even further, we can say that if such conceptual constructs as mythological stories or mathematics were invented to create meaning and sense in our concepts, paintings and other artworks were created to find logic and coherence in our visual perception.

It is not so surprising therefore that much of what we consciously know about visual space, about shapes, colors, and spatial relationships, was actually learned from drawings or paintings. All of us, to take a simple example, learn intuitively early in life, that if there are two objects in our line of sight, the nearer one will block out parts of the more distant one. But we only became consciously aware of the rules underlying this phenomenon when we first attempt to draw a picture involving such objects. The eye, in other words, can spontaneously take in the visual environment. This ability is built into the unimpaired nervous system. But this is not the same as conscious visual comprehension. The difference between the two is essentially the same as that between genetic and symbolic learning. Or, to return to an earlier example, it is like the difference between being able to catch a ball on the one hand and being able to calculate its path on the other. Many of us, of course, never bother to learn how to calculate mechanical problems because we do not need to understand mechanics to make practical use of its laws in everyday life. Similarly since most of us can see easily, we seldom consciously investigate and explore our visual environment.*

For a painting to be able to represent the three-dimensional

* Only a minority seem to care about the active, conscious use of vision. These are mainly people with a strong interest, active or passive, in the visual arts. The majority of us, when not reading, do not use vision more than is necessary for physical and social survival. We use our eyes, in other words, for orientation and recognition, but seldom for exploration. In our society, furthermore, the conscious use of eyes and of visual images for searching and learning has never been considered important enough to be taught in schools in any serious way. It is neglected even in the teaching of the visual arts. Art history texts or art appreciation classes often put excessive emphasis on social and psychological factors in the lives of artists, neglecting the visual analysis of their works and thus contributing little to the learning of the active use of vision.

world, it has to contain certain visual distance cues; without them it will not create spatial illusion. The creation of such cues is not simple. In fact, *successful* attempts at producing optically realistic pictures in a systematic way are relatively recent. The first such pictures probably appeared during the Hellenistic-Roman era, or about two thousand years ago. Interestingly, the realistic style of many of these paintings was inspired by their use as theatrical scenery. Other examples of early realistic paintings have also survived—for example, on some of the walls of the houses of Pompeii. But with the decline of the ancient civilizations, the intuitive knowledge of realistic painting vanished completely from consciousness. As in science and mathematics, almost a millenium and a half had passed before European civilization first reached the level of these late Classical achievements in the pictorial representation of three-dimensional realism.

Yet as slow as they were in catching up, the arts were still way ahead of the sciences. At the beginning of the fourteenth century, for example, when to most philosophers of nature the idea of observing nature for the purpose of learning was entirely alien, certain Italian painters—Giotto among the most important and certainly the most admired—were using a detailed observation of nature as the key to visual realism.* Around the end of the fourteenth century, similar ideas also became accepted in the Netherlands, and there, painters such as Jan Van Eyck also created startlingly realistic pictures. None of these painters seems to have had a technical basis for their realism. They were just careful observers who attempted with incredibly detailed accuracy to recreate the optical appearance of depth and movement and the distribution of figures in space. They were successful because the coordination between their hands and their eyes was extremely effective. Giorgio Vasari, the fifteenth-century chronicler of the lives of the artists, could have written about each of these artists what he wrote about Giotto: that he was "not so much the pupil of any human master as of Nature herself. . . ."

* Certain isolated thinkers did advocate even earlier that knowledge of the world must be based on observation and not on authority. The best known among them was the thirteenth-century English Franciscan friar Roger Bacon. His contemporaries, however, were not impressed. All Bacon got for his ideas was a long stretch in prison.

But in the first two decades of the fifteenth century, two important visual cues used in three-dimensional space perception—the diminishing of the size of objects with distance and the visual convergence of receding parallel lines—were beginning to be reproduced on flat, two-dimensional surfaces. The practical use of the laws of linear perspective had finally been established. The method for making practical use of these laws was first formulated in the early years of the fifteenth century by the Florentine architect-painter-scholar Filippo Brunelleschi and fully developed not long afterwards by his colleague and compatriot Leon Battista Alberti.

The practical way of making a perspective drawing was first described by Leonardo da Vinci essentially as follows: Put a pane of glass in front of you perpendicular to your (imaginary) line of sight. (Use one eye only to avoid corrections of binocular vision.) Now draw the outlines of the objects appearing through the glass by tracing them exactly as they appear through the glass. This drawing would be, by definition, constructed according to the principles of linear perspective. This simple exercise allows you to depict the visual organization of space exactly as the eye sees it and thus to produce realistic pictures. Half a century after Leonardo, the German painter Albrecht Dürer depicted how Leonardo's prescriptions can be put into practice [Figure 6-1].

This type of optical realism can be achieved on canvas or paper if we observe the laws of simple (one-point) linear perspective: (1) Images diminish in size in a well-defined way as they recede from us; (2) receding parallel lines appear to converge at a

Figure 6-1

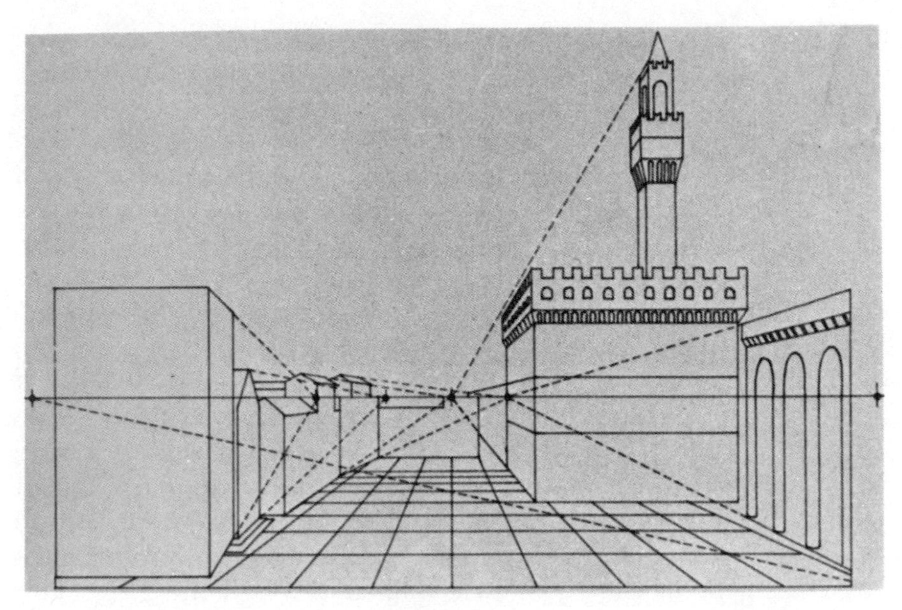

Figure 6-2

point (called the vanishing point)* on the horizon line (the ideal line of the horizon as it would appear in a perfectly flat environment through the pane of glass); (3) receding lines seem to go down if they begin above eye level and seem to go up if they begin below it; and, (4) just as in everyday experience, objects close to the viewer overlap those farther away on the same line of vision. By consistently applying these rules, the artist can reproduce those visual cues which are determined mostly by the idealized optical-geometrical structure of the human eye, and thus give a good illusion of a three-dimensional, homogenous, infinite space on a two-dimensional flat surface. (The drawing in Figure 6-2 shows a schematic reconstruction of Brunelleschi's lost perspective panel of the Piazza Della Signoria in Florence [after D. Gioseffi].)

Simple as these laws are, their application in the visual arts profoundly transformed the arts themselves, led to a new understanding of human vision as a space-exploring sense, and brought into existence a new image of space, a symbolic space which was

* The position of this point is determined by the fixed position of the observer's eye.

Figure 6-3

destined to dominate the spatial conceptions of Western civilization for the next five centuries.

One of the first examples—and a beautiful and instructive one at that—of the use of linear (one-point) perspective is Donatello's *Feast of Herod,* which dates from about 1425 (see Figure 6-3). The work is not a painting but a bas-relief sculpture; it is, however, almost two-dimensional and thus at least optically similar to a painting. It was made by the Florentine sculptor for the famous baptismal font in the baptistery of the Siena Cathedral.

Donatello sculpted the *Feast of Herod* about ten years before Alberti wrote his description of the theory and praxis of perspective painting. But it applies the theory so perfectly that the art historian Frederick Hartt actually suggested in his recent Donatello monograph that this artist "may be the true originator of the perspective promulgated a decade later by Alberti." So masterful is the application of the rules of perspective that the viewer's eye is literally led, as Hartt puts it, "through arches into a second hall, through more arches into an inner chamber, and at the far right under an architrave into an ascending staircase," the artist thus creating a realistic representation of an extensive indoor space on a surface not much bigger than a breakfast tray.

This new notion of space seems to have had a significant impact on other aspects of the visual arts. Like all the sculptures of Donatello, the *Feast of Herod* broke not only with medieval technique but also with the medieval outlook. For example, medieval visual conventions would have demanded that the dominant person or events occupy the center of a picture. The center of this relief, in contrast, is almost empty. This emptiness, however, communicates visually a meaning which we understand even if we are not conscious of our own understanding. The empty middle, as P. Janson of New York University points out, first of all "conveys, more effectively than the witnesses' gestures and expressions, the impact of the shocking sight [of St. John's severed head]." In addition, "The centrifugal movement of the figures helps persuade us that the picture space does not end within the panel but continues indefinitely in every direction; that the frame is merely a window through which we see this particular segment of unlimited, continuous reality."* The relief not only created a more realistic visual image of space than did earlier works but also conveyed human emotions such as the feelings generated by the horrible sight of the severed head of St. John the Baptist more directly. These emotions, no less than the sight of extended space, were visible and were communicable to any viewer. No longer was it necessary for the viewer to know the story in order to understand the message. The message lay in what was depicted, in the visual.

* Another pioneer artist who used perspective as a basic tool of his art was the Florentine painter Masaccio. He was twenty-four and had only three more years to live when in 1425 he painted the fresco entitled *Tribute Money* that is now in the Branacci Chapel in Florence. This was probably the first painting to depict outdoor space realistically and to show some of the effects of the play of light and shade. A couple of years earlier Masaccio had painted the *Trinity* fresco in the church of Santa Maria Novella in Florence. This fresco appears to be the first surviving painting constructed using the laws of linear perspective. Earlier paintings by Brunelleschi that use linear perspective are lost.

It was, of course, not the perspective method per se which created great art. The method was, after all, an ingenious technique, a prescription, which in merely competent hands could produce realistic effects but nothing more. But in the hands of the great artists it became the means to "recreate reality in a way that is convincing to the eye as well as to the mind," as the art historian John White wrote. To have created harmony between the eye and the mind, to have established a correspondence between perception and symbolic representation, was one of the main accomplishments of the artists of the Renaissance.

The success of the earliest Renaissance works inspired further studies both in the theory and in the praxis of perspective. Among the first artists to write learned treatises on the subject were the painters Paolo Uccello, Piero della Francesca, and Leonardo da Vinci.* There were, of course, other great artists who, although not very interested in the problems and the rules of the theory, did exploit them to create their own distinctive art, and they were also very successful. We see now in hindsight that the development of the visual arts was by far the most important mental activity during the fifteenth and sixteenth centuries. At a time when neither science nor philosophy was able to offer any new and exciting ideas about the world, such artists as Fra Angelico and Ghiberti, Mantegna, Perugino, and Raphael, Titian and Tintoretto, and the greatest of all, Michelangelo, were able to define, ask, and answer the important, if unexpressed, questions of the age: What do we see when we see? How do we learn when we see? What can vision say about the world and about the human role in it? What are the visual aspects of the important human feelings, concerns, and values, and how can these be expressed?

There are probably more distance cues in human vision than we are conscious of. The ones we know about range from the trivial (objects obscure anything further in the line of vision) to the subtle (the changing of hue with distance). But only the two basic cues which are used in linear perspective have the distinct

* The picture "Perspective View of an Ideal Town" [Figure 6-4] is attributed to Piero della Francesca. It depicts a Renaissance ideal while showing the power of the perspective method. The painting is in Urbino.

Figure 6-4

property of obeying mathematical laws. They also appear to be the simplest precisely because they are determined by the idealized geometric-optical properties of the eye. These cues can, after all, be reproduced quite faithfully by something as simple as a camera. Many or perhaps most other distance cues are far more sophisticated, and our use of them seems to have evolved through some probably very complex mechanisms in our visual information processing. At least some of these other cues should therefore be generally subtler, more sophisticated, and far less obvious than those of geometric perspective. Yet at various times throughout history painters in various civilizations had discovered and made use of many of these more-sophisticated cues, while the visual cues of linear perspective—which common sense tells us should be the most universal, the most objective, and therefore the least culture-dependent—were introduced into painting once only.* This is a remarkable fact, especially if we take into account the fact that the theoretical laws of perspective were discovered long before the Renaissance. The theory was known both to the Greeks and Romans; its geometric foundations were already clearly described in Euclid's *Optics.* It was disseminated in Europe in the Middle Ages through the works of the Arab scholar Alhazen. Yet it did not occur to the Greeks, to the Arabs, or to the medieval Europeans to make use of these laws in painting.

But when, in the "quattrocento" (the term used to refer to the Italian civilization of the fifteenth century), the laws of linear perspective were put into practice, the effects reached far beyond the artistic sphere. The invention of this method, the unprece-

* It has been debated repeatedly whether perception of the cues of perspective is indeed universal. The noted art historian Erwin Panofsky once characterized linear perspective as being essentially a convention, a symbolic form among the many used to represent space. Most art historians and perceptual psychologists seem to disagree with Panofsky and claim that we actually see the world in perspective. This view is supported by the obvious fact that there are numerous features of the perspective method which generally correspond to the optical workings of the human (and even nonhuman) eye as well as the camera. Panofsky's viewpoint, on the other hand, is supported by observations which seem to show that people unfamiliar with Western painting and photography are not able to comprehend perspective information. These contradictory findings might be another example confirming that in seeing, as in many other human functions, we have to learn and to nourish even what is given to us by nature.

dented proliferation of great artists and great works of art over a relatively short period, and the central role of the visual arts in the mental life of the fifteenth and sixteenth centuries were all manifestations of the same process: a reevaluation of the role of vision in human cosmology. The introduction and widespread use of linear perspective amounted to nothing less than the emancipation of human sight from the prejudices of philosophers. Many of the most influential philosophers from Plato onwards had emphasized that the human senses are imperfect and therefore cannot convey reliable information about the world. But if it were possible to include the sense of sight among the rational faculties of the mind, Plato and the others could be proved wrong. The arts were giving visible proof that as well as taking in an infinite variety of often random visual sensations and organizing them into useful patterns, the human eye also obeys simple, yet exact laws.

Thus the widespread use of geometric perspective technique in the arts became a new source of and a support to the emerging Western intuition that the rationality which could be created by the human mind in the realm of concepts could also be created in the world of sensory perceptions, and that in order to make the world rational and orderly, we have to first make our sensory perceptions rational and orderly. This was the same intuition which had led earlier to the evolution of exact measures of time in polyphony, and that was to lead in another two centuries to the appearance of the experimental sciences.

The Renaissance was, then, the epoch when the conscious and systematic examination of *space as a source of human sensory experiences* commenced. The emerging space of Renaissance art opened up a new symbolic world which could be manipulated and investigated and in which new, faithful, and extremely rich models of reality could be constructed and studied. This new space, the like of which had never been seen before, fascinated its public the same way that the space opened up by the telescope and microscope fascinated their great-grandchildren a century later, and that the new discoveries about cosmic space fascinate us. Unlike the artists of earlier and later times, the artists of the Renaissance were quite conscious of the great role their art played in the emerging passion for learning by observation. "If you disparage painting," wrote the Renaissance genius Leonardo da Vinci, "which alone imitates all the visible works of nature, you disparage a most subtle science which by philosophical reasoning examines all kinds of forms."

The new outlook was manifested in all aspects of the visual arts, in sculpture as well as architecture. After a fifteen-hundred-year hiatus, during which statues had always clung cautiously to the wall, never daring to move, freestanding sculptures depicting movement had already begun to reappear early in the Renaissance. In the buildings of the great Renaissance architects, Michelozzo, Bramante, and Palladio, the self-defeating complexity of the Gothic architecture of the late Middle Ages gave way to simple lines, to mathematical proportions, and to Classical spaciousness.

The new visual space of the Renaissance changed not only the way people felt about space, but also the way they thought about it. The calm, neutral, and organized space of this era was no longer based on the imaginary symbols and values of the supernatural but on the measurable and mathematically describable rules of visual perception. The important points and regions in space were those of visual, rather than religious, significance. Size now indicated distance rather than religious or secular rank. Spatial allegory took second place to visual reality. In painting, the vanishing points and horizon lines represented the realistic limits of visual perception and not the limits of space and the human world. The "left" was no longer worse than the "right" as both were equally visible. Even "up" was not necessarily superior to "down"; their respective merits were subject to discussion. This is one of the subtle messages in Raphael's famous *School of Athens* [Figure 6-5]. The central figures of Plato and Aristotle are discussing the merits of observing the world down here (Aristotle pointing downwards) or of trusting the world of Plato's ideas up above. The fresco (located in the Vatican) is also a beautiful example of the use of perspective to evoke a grandiose, large-scale space on a relatively small surface. Space could, in other words, now be viewed as homogenous and isotropic with no preferred places and no preferred directions.

In discussing the evolution of the new notions of space, John White remarked that in pre-Renaissance paintings, "It was possible to see space gradually extending outwards from the nucleus of the individual solid object. . . ." In the Renaissance, however,

Figure 6-5

"Space is created first and then the solid objects of the pictured world are arranged within it. . . ." Or, as White put it even more succinctly: "Space now contains the objects by which formerly it was created." As a result, space was no longer influenced by what it contained, it was unchangeable and indestructible, it was always there whether empty or not, it was measurable, and it evoked a sense of expansion, of distance, and of infinity. It evoked, in short, the features of sensory perception which were to be conceptualized later in physics as "absolute" space.

A space which was explorable by the senses and was not haunted by spirits and assorted supernatural monsters naturally invited travel. The Renaissance was also the time of the beginning of the great explorations. The sense of a rational and measurable space encouraged the drawing of the first maps that were not merely allegorical but actually attempted to represent the real world in a rational symbolic mode. The very map which led Columbus halfway around the world was made in Renaissance Florence by the famous scholar Paolo Toscanelli, a friend of Brunelleschi and Donatello.

Geometry, that other great legacy of Greek symbolic space, was also reborn. Euclid, Appolonius, and Archimedes were studied with as much fervor as the Greek philosophers and Greek art. Some of the knowledge that had been lost during the Middle Ages was revived in this era; the rediscovered knowledge included trigonometry, the study of triangles. Regiomontanus, a German mathematician of the fifteenth century, traveled to and studied in Italy, where he became acquainted with the Greek and Arab works on trigonometry. He reconstructed, developed, and published these studies, and trigonometry soon became a powerful tool in pure mathematics as well as in the sciences and engineering.

But it was only long after the Renaissance that the Greek standard of accomplishment in geometry was finally and decisively surpassed. The geometry of visual space, on which the theory of linear perspective was based, was formulated with great generality and mathematical precision by Girard Desargues in 1639. As Euclid had developed a geometry to fit a tactile space, a space where parallels never meet, where a circle remains a circle no matter how you trace its line with your finger, so Desargues created a geometry for a visual space where parallels do meet at a point, where a circle will look like an ellipse or even like a line when viewed obliquely. Desargues' work became known as projective geometry. It is now an important and beautiful part of both pure and applied mathematics.*

Simultaneously with Desargues' work, another and perhaps even more important step in the development of geometry was taken by two part-time mathematicians: René Descartes, the great philosopher; and Pierre de Fermat, a lawyer, a civil servant, and incidentally a most important mathematician. Quite independently of each other, Descartes and Fermat discovered how two seemingly entirely different realms of thought, numbers and equations on the one hand and geometrical figures on the other, could be unified in a single beautiful framework. This framework became known as analytic geometry. The new method amounted

* Desargues' contemporaries were, however, not impressed by his system. The only important mathematician to work on projective geometry in Desargues' time was Blaise Pascal, but he soon abandoned it when he turned to theology. Desargues' work was forgotten until Jean Victor Poncelet, an officer in the army of Napoleon, rediscovered and developed it in 1813 while a prisoner of war in Russia.

to an abstract mathematical characterization of spatial (two- and three-dimensional) figures, to a unification of the Greek legacy of geometry and the Greek-Hindu-Arab legacy of algebra. It translated geometry to numbers, figures to concepts; it gave spatial figures numerical interpretation, equations spatial representation; and it paved the way for the discovery of calculus and for the creation of an immense variety of symbolic spaces both in physics and in mathematics. As Descartes himself put it: "I believed that I could borrow all that was best both in geometrical analysis and in algebra, and correct all the defects of the one by the help of the other." With the discovery of analytic geometry, European mathematics not only surpassed the Greeks but also broke free from all traditions and embarked on an entirely new and independent course. It was to become one of the most imaginative and successful enterprises the human mind has ever designed.

The search for order and mathematical proportion in sensory experience, which had started in polyphonic music and was further developed in Renaissance painting, culminated in the seventeenth century in the evolution of the experimental-mathematical sciences.* At the beginning of the century, Galileo was the first to successfully employ sensory observation to find the mathematical laws governing not just patterns in space or in time but also in the combination of the two, i.e., in the more complicated process of motion. He was interested, in particular, in that most common and ordinary motion: the motion of free-fall. About thirty years after Galileo's death, Isaac Newton brought together the Copernican-Keplerian astronomy with Galileo's laws of motion and unified them into a single magnificent framework. He discovered that what Copernicus and Kepler had learned about the solar system and what Galileo had found out about the laws of free-fall could be seen as the manifestations of a single fundamental law of nature—the law of universal gravitation. He realized that gravitation was the one force that holds us to the earth, holds the moon around the earth, and keeps all the planets revolving around the sun. In doing so, he literally unified the hitherto separate worlds

* To avoid misunderstanding, I mention here again that the word "science" in this book always denotes the fundamental natural sciences only.

of the sky and earth and, for the first time in history, demystified the heavens. In addition, Newton was able to solve the general problem of motion (not just the problem of free-fall) in a final and perfect form.

Newton also produced clear conceptual formulations of the scientific ideas of time and space. I have already quoted his description of absolute, independent, "equably flowing," measurable time. His definition of absolute space was just as firm and unequivocal. He started by emphasizing the independence of space from every other aspect of nature: "Absolute space, in its own nature, without relation to anything external, remains always similar and immovable." These definitions more than anything else established the conceptual framework of absolute (also sometimes called Newtonian or, as here, classical) time and space and helped people become fully conscious of the nature of their new universe.

From this point on, science increasingly determined what people thought and even what they felt about time and space. In less than two hundred years, science had changed the general conception of the nature of the universe and of the position of humans in it more fundamentally than anything else before or since. In the year 1500, highly educated Europeans were still taught to believe that the universe was centered around the earth, that it was finite in extent and surrounded by netherworlds and populated by superhuman creatures like angels and devils. By 1700, the same class of people—religious or not—knew that they were living on a small planet orbiting the sun in a universe which was infinite in extension and yet had no longer any place for netherworlds.

The transition from the old to the new cosmology was a quick one; still, people's beliefs did not change all at once. Kepler, who discovered (in the first years of the seventeenth century) that the planets revolve around the sun in elliptical orbits, thought that they did so because angels were pushing and directing them. Two generations later, in creating the science of classical mechanics, the science of motion, Newton replaced the angels with the mathematical laws of gravitation and inertia. These turned out not only to underlie Kepler's discovery but also to explain all the observed phenomena of the mechanics of the solar system. From then on, neither angels nor any of the other colorful symbols of earlier cosmologies could play a role in the new universe of science.

What remained after the old ideas and symbols were removed was the simple universe of classical mechanics. A world which consisted of *permanent objects in space and time* obeying the mathematical laws of inertia and gravitation. Historians of science, for obvious reasons, like to call this model the "mechanical universe." From an evolutionary perspective, however, one could just as well call it a mammalian cosmology with a human face—a recreation of the built-in perceptual cosmology of the human brain within a mathematical framework. For this mechanical universe is the symbolic cosmology which most closely approximated the one built on the basic mammalian perceptions.

We know that our brain evolved in such a way that it could only meaningfully process information from direct sensory data by building a world of space and time and objects in space and time. The early scientific was the first symbolic cosmology in history in which primary sensory data were considered the main source of information and in which most concepts reflected something which was directly or indirectly perceivable.* In all earlier, nonscientific, traditional or mythological cosmologies, the world was viewed as populated with the gods and spirits associated with them, and none of this was ever supposed to be perceivable. As a result, the importance of sensory input, of observations of the world of ordinary objects in time and space, was either ignored or else the senses were downgraded as unimportant or unreliable. Therefore, these earlier cosmologies referred only vaguely and haphazardly to ordinarily perceivable phenomena, and their worldviews were usually far removed from the sensory experiences of the brain.

The early scientific cosmology with its emphasis on sensory data was very much a mammalian one but, of course, in a very special human sense.† Human cosmologies, scientific or not, al-

* This property made the scientific the first human cosmology open and understandable to anybody. I mean this in the sense that the scientific worldview is much less dependent on particular social and cultural values than any of the prescientific or traditional cosmologies. The human senses and the basic laws of mathematics are the same all over the world.

† By human I do not mean to imply humanistic. Mathematics, for example, is a human creation although it does not deal in human values.

ways consist of ideas and not of sensory data. Thus even though the world of early science resembled the world of mammals in the sense that it consisted of permanent objects moving in space and time, it was nevertheless an abstract model. It was built on abstract ideas of time and space which evolved from but did not accurately mirror our actual experience.

One assumption was, as mentioned, that both time and space were measurable exactly and reliably by the senses. But there were other no less important ideas. Both time and space were, for example, supposed to be homogenous and uniform. This meant that no domain in space was different from any other: the space that was "here" had the same properties as the space that was "there." Space was one and the same thing all over the universe, and therefore the space of science could contain no holy, charmed, or tribal places. Uniformity was also thought to characterize the flow of time. As time was passing now, so it would pass in the future, however distant. Nor had the passage of time been any different in the past, however remote. And no segment of time was more important than any other segment. The flow of time did not reflect holy days, anniversaries, and the like. Space was also supposed to be isotropic: no direction was any better or any worse than any other direction. Time, in contrast, was not isotropic but passed always in one direction only. Time had an arrow which pointed from the past to the future through the "now" which was the only perceivable segment of it. The essential difference between the past and the future was that one could influence the latter but not the former.

In contrast to ever-flowing, dynamic time, classical space was static. It did nothing. It merely existed and contained in itself everything in the world. Since it was measurable, it possessed a geometry. It was assumed that this geometry was Euclidian; that the theorems and rules of Euclidian geometry were valid everywhere in the universe. And "everywhere" meant a lot since space was supposed to be infinite. Time also was infinite. In the classical picture, neither time nor space could possibly have a beginning, an end, or any kind of boundary.

Both time and space were assumed to be continuous. No matter how small a segment of either you imagined, it would still remain "time" or "space." The continuity of time and space was reflected in the continuity of motion. When getting from a point

A to point B on a certain path, a body traversed all parts of the path but nothing else. Other changes were also assumed to be continuous. When a body cooled from, say, a temperature of 30° to 20° (in whatever units), then the body was, at a certain time, necessarily at a temperature of 25.188°.

In the classical view, furthermore, time and space had absolutely nothing to do with each other; they were completely separate and independent entities. They might be connected in our minds when, for example, in describing motion, we construct such concepts as velocity or acceleration. But these are our mental constructs only. "Out there," in reality, space is space and time is time and no intrinsic connection can ever be imagined between them.

And last but not least, classical time and space were both absolute. Not only did they exist independently of us, of our perceptions, and of each other, but they were completely independent of anything else in the world. That meant that nothing whatever could possibly modify the objective properties of space or the flow of time. The distance between two points, for example, was defined by these points once and for all, no matter when, why, by what, and under what circumstances it was measured. The same was true about the time interval between two events. Time and space existed independently of any kind of matter and its motion. Matter could exist in space and time only. But time would continue to flow imperturbably and space would still spread out to infinity whether the universe contained any matter or not.

The influence of these ideas is still strong and, to many of us, they still sound natural and self-evident. But, in fact, there is little that is actually natural about them. Although they evolved from direct sensory perceptions, they were abstracted from them, to such an extent that they often contradicted what we spontaneously perceive. No earthbound creature, to take the most obvious example, can ever perceive space as isotropic. The difference between "up" and "down" is very real, and taking into account such differences is a given of our neural makeup. Isotropy, therefore, is an abstraction and a difficult one at that. Learned professors of European universities argued for centuries during and after the Middle Ages that if the earth were round, the living creatures on its "other" side would fall "down" into the void. Nor do we ever feel that space, to give another example, is homogenous. We are our own reference system always and this gives "our" place a standing different from any other place in the universe.

The scientific notions of time and space were terrifying to many people. In removing all the symbols of subjective values and concerns from it, science created the first human cosmology which was indifferent not only to everyday human fears and hopes but even to the very existence of the human race. No wonder that the emergence of such a view of the world produced fears and anxieties. An early testimony by the seventeenth century mathematician-mystic Blaise Pascal reflects these anxieties movingly:

> I see the terrifying immensity of the universe which surrounds me, and find myself limited to one corner of this vast expanse, without knowing why I am set down here rather than elsewhere, nor why the brief period appointed for my life is assigned to me at this moment rather than another in all eternity that has gone before and will come after me. On all sides I behold nothing but infinity in which I am a mere atom. . . .

Pascal was neither the only nor the last person to feel alienated by this scientific worldview. His fears and apprehensions, the trauma caused by the scientific worldview, were frequently echoed in later centuries. But others, among them many active scientists, took a different view. In the success of science they saw the comforting proof of their belief in a universe that was rational and not chaotic, and law-abiding rather than arbitrary. The need for this belief originated in the Judeo-Christian doctrine of a single Creator whose laws are rational and thus accessible to human reason. Plato's teachings about the perfect ideas underlying an imperfect world of the senses were also influential. Whatever its origin, an a priori belief in the orderliness of nature was necessary for the evolution of science. It would have been pointless to use sensory observation to find mathematical laws in nature without assuming that such laws existed in the first place. And it was satisfying to many scientists that these assumptions were vindicated by the authority of science itself.

The supreme principle which assured rationality and orderliness was *causality*. If an event was inevitably followed by some other event, we acknowledged the observed inevitability by calling the first event the cause of the second. The perception of causality, like that of space and time, is probably part of our mental equipment since the recognition that some events are always or very likely to be followed by certain others seems to be a key factor in adaptation. Adaptation cannot take place in a wholly acausal world—one where events have neither causes nor consequences.

With the scientific revolution, causality became not just the dominant but virtually the only ordering principle of the external world. And with reason too. The laws of classical mechanics were just the type of laws which one would expect from a perfectly causal world. According to Newton's laws of motion, if you knew all the forces acting on a body and you also knew the position and the velocity of the body at a certain instant, you could then calculate as accurately as you wished where the body would be and how it would move at any other instant in the future.* Such calculations were indeed made with respect to the motion of planets and comets (for example, Halley's exact prediction of the return of "his" comet), and the accuracy of such calculations impressed the thinking public tremendously and contributed greatly to the establishment of the general worldview of a fully mechanistic, causal universe.

The predictive power of classical mechanics was quite without parallel in any previous human experience. The source of this power and the most important tool in its use was the mathematical method known as calculus. Calculus, coeval with classical mechanics,† provided the perfect machinery for describing smooth motion and continuous, predictable changes. It was the ideal mathematical method for portraying the past, present, and future in causal terms. In contrast, it was, at least initially, unable to deal with sudden, unexpected changes, or with jerky, discontinuous motion. In other words, it fit right in with the world of classical science, with the mammalian model of the world, in which the

* From the same data you would be able to calculate its past history as well.

† The method was discovered simultaneously and independently by Newton and the German mathematician-philosopher Leibniz.

brain smoothed out and made continuous and predictable the essentially jerky, disconnected stream of incoming information.

So impressive were the accomplishments of calculus-based classical mechanics that by the late eighteenth century, some physicists actually thought that the world contained nothing else but simple bodies (atoms) influencing one another by their mutual gravitational attraction and following their prescribed continuous paths according to the equations of motion. An extreme formulation of this view was made by the French mathematician, Pierre Simon Laplace. He actually thought that for a hypothetical superior intelligence (commonly referred to later as Laplace's demon) who would know all the forces acting in the universe and the positions and velocities of all the atoms at a given instant, there could be no surprises. The knowledge of these data would enable the demon to determine what the motion of every single atom at any future time would be.* While Laplace himself did not believe that such an intelligence could exist, he added nevertheless with the characteristic self-confidence of classical science that in achieving such a perfect creation as the science of astronomy, the human mind had already given a modest example of such an intelligence.

Returning to the scientific problems of space and time, we shall see that there were some rather compelling reasons to view both time and space as absolute. The most important ones were connected with the very laws of classical mechanics. Consider the first law, which originated essentially with Galileo. It deals with one of the simplest and seemingly one of the least exciting phenomena one could imagine: the movement of a body on a straight line with unchanging velocity. In Newton's formulation the law

* This philosophical interpretation of classical mechanics was, of course, neither the only nor the first strongly fatalistic worldview. Many of the world's religions were, and some still are, just as fatalistic. About seven hundred years before Laplace, for example, the Persian poet Omar Khayyam expressed Laplace's fatalistic philosophy perhaps less scientifically but certainly not less powerfully:

With Earth's first Clay They did the Last Man's knead,
And then of the Last Harvest sow'd the Seed:
Yea, the first Morning of Creation wrote
What the Last Dawn of Reckoning shall read.

reads: "Every body continues in its state of rest, or of uniform motion in a right line, unless it is compelled to change that state by forces impressed upon it." This is an exceedingly interesting law for several reasons. In a historical context, it represented a revolutionary new view of motion. Before Galileo, it was thought (mainly based on the faulty but universally accepted physics of Aristotle) that an ordinary object had only one natural state and that was rest. Rest was natural since a body would stay at rest until something moved it. Motion, on the other hand, was not natural since a body would not move unless something, some kind of a mover, forced it to move. When the action of the mover stopped, the body would stop moving and would again return to its natural state of rest.*

Galileo and Newton challenged this view and made uniform motion on a straight line (called "inertial motion") as natural a state for a body as rest. A body will continue to move, they said, not as long as something *moves* it but until something *stops* it. This is essentially the law of inertia. It is an abstract law and does not follow from superficial observation. In fact, we never actually see it satisfied. None of the bodies we see will move uniformly in a straight line forever. All will sooner or later come to a stop. But we know now that they do so because external forces (such as friction and air resistance) stop them.

It was precisely this law of inertia which required, according to Newton, that space be absolute; absolute space was necessary because the law required an absolute reference system. For if a body were to continue moving forever on a straight line and with uniform velocity, then something had to exist that determined what was uniform and what was straight. There had to be a fundamental property of nature, in other words, relative to which motion could be uniform and straight. This fundamental property, according to Newton, was space itself, which included explicitly or implicitly an absolute reference system: "That the centre of the system of the world is immovable. This is acknowledged by all. . . ." A body either moved relative to this center and then it was really moving, or it did not and then it was at rest. Motion, like space and time, was absolute.

* Recent studies show that a sizeable fraction of educated people still believe that these ideas are essentially correct.

All this was relatively clear and straightforward. Problems started when one wanted to determine the framework of absolute space. Then the laws of mechanics played a joke on their inventors. For such a determination turned out to be impossible in the framework of classical mechanics. Imagine two systems: One is at rest in absolute space, the other is moving on a straight line with uniform speed relative to it. The two look as if they move relative to each other but only one of them really moves, i.e., moves relative to space, and you would like to find out which one. Yet if you perform mechanical experiments in both systems, the results will be exactly the same in both. We know this by now from ordinary experience. On a smooth-flying aircraft, we do not feel the motion at all and we can do all sorts of mechanical experiments with free-fall or with a pendulum or with colliding balls or with any other mechanical device, and the experiments will all show the same results as if they were performed in a laboratory at rest relative to the earth. There are no mechanical experiments which could distinguish between systems which move uniformly and on a straight line relative to each other. All such systems are equivalent for mechanics. This fact was already known to Galileo and is now called the Galilean principle of relativity.* This principle not

* This principle was described by Galileo with a force and eloquence which is missing from the dry and cautious language of contemporary science. Here is how Galileo argued:

> Shut yourself and a friend below deck in the largest room of a great ship, and have there some flies, butterflies, and similar small flying animals; take along also a large vessel of water with little fish inside it; fit up also a tall vase that shall drip water into another narrow-necked receptacle below. Now, with the ship at rest, observe diligently how those little flying animals go in all directions; you will see the fish wandering indifferently to every part of the vessel, and the falling drops will enter into the receptacle placed below. . . . When you have observed these things, set the ship moving with any speed you like (so long as the motion is uniform and not variable); you will perceive not the slightest change in any of the things named, nor will you be able to determine whether the ship moves or stands still by events pertaining to your person. . . . And if you should ask me the reason for all these effects, I shall tell you now: "Because the general motion of the ship is communicated to the air and everything else contained in it, and is not contrary to their natural tendencies, but is indelibly conserved in them." And at another time you shall hear the specific replies and a full explanation.

only followed from practical observations but was a strict mathematical consequence of the laws of mechanics. Thus those very laws which seemed to have demanded the existence of absolute and immovable space in the first place themselves forbade the detection of it.

If one apparently cannot ever prove the existence of absolute space, then why even talk about it? However, there were other phenomena which seemed to prove that absolute space was, after all, not a fiction. If, for example, our motion is not uniform and straight but accelerating (or slowing down), then there is no question of not knowing whether we move or not. We can feel the effects of acceleration very well as do all instruments. It can easily kill us as it sometimes does in car accidents. There seems to be nothing relative about acceleration, and it was natural to assume that acceleration was indeed an absolute process, that it was acceleration against absolute space. It was therefore felt that the impossibility of detecting absolute motion in systems which move uniformly on a straight line, while confusing, was of no fundamental significance. And it was also hoped that at some point later on, new kinds of experiments, nonmechanical types possibly, would be able to detect and prove the absoluteness of space and motion.

It was mainly for such reasons that the existence of absolute space was universally accepted. Soon it became an axiom whose validity was generally assumed and no longer questioned. And as far as time was concerned, the fundamental properties of absolute time were even more firmly accepted as being real than those of absolute space. No physicist questioned, for example, that time passed uniformly, or that it flowed at the same rate always and everywhere, in spite of the fact that the idea was, in fact, quite absurd. Against what can one measure, after all, the flow of time to see whether it was uniform or not? Nevertheless the idea of absolute time flourished, and not until Einstein came on the scene after almost three triumphant centuries of classical physics was its reality seriously questioned.

Besides the axioms of absolute time, space, and motion, there evolved slowly but irresistibly another belief which also became something of an axiom. This was the belief that the universe was filled with a curious type of matter called the "ether." This was as little doubted by physicists before the twentieth century as was the "even rate" of the flow of time or the "immovable center" of space. And as with absolute time and space, there were some good reasons why the existence of the ether was so widely accepted.

Who needed the ether and for what purpose? Consider the law of gravitation as discovered by Newton. It says that any two masses anywhere in the universe would always attract each other with a force which is proportional to their masses and inversely proportional to the square of their distances. According to this law, the sun and the earth would be continuously attracted to each other by this force. No problems so far. But if we would like to understand this law a little better, we should consider it in action. Force is something physical, it has to be "felt" by a body if that body is to be influenced. If I am pushed or pulled by somebody or by an object, I feel the force. Now how, where, and by what means does, for example, the earth feel the attraction of the sun? What, in other words, tells the earth that it is being attracted by a body a hundred million miles away? The sun does not seem to send out arms or ropes to pull the earth toward itself. How does the earth "learn" about the attraction? What physical agent communicates it?

This is a serious question.* The only answer imaginable at the time was that there must be some kind of matter between the sun and the earth and thus, by implication, everywhere in the universe (since two bodies would attract each other anywhere), which would actually transmit such forces. This matter, which was called the "ether," had to have curious properties. It had to be invisible and, if a body moved in it, it had to offer no appreciable resistance.† Yet its consistency had to be strong enough to transmit the tremendous gravitational pull of the sun.

Still, the ether was a rational, acceptable explanation even if nobody knew how to "catch" it or how to prove that it really existed. What made the existence of ether even more logical was that not only did the force of gravitation demand it, but so did the phenomenon of light. In the end it was these concerns about the propagation of light more than anything else which secured the

* The problem of agents which communicate forces is still an important issue in contemporary physics.

† Even a small resistance would, in time, slow down the motion of all the planets in the solar system, and there was no evidence showing such a slowing down.

ether such a prominent place in classical physics that its reality was not seriously challenged until the twentieth century.

There were in the seventeenth century two competing ideas concerning the nature of light. Newton suggested that light consisted of small particles traveling in a straight line, which is a logical assumption since we often see light propagating in this way. The Dutch physicist Christian Huygens, however, had another idea. His starting point was the observation that light from a source propagates in all directions, and in this property he saw an analogy with waves. Water waves, for example, also propagate in all directions from their source—think, for instance, of the ripples caused by a thrown stone. Huygens' ideas turned out to be quite correct, and during the 1700s and 1800s, it was shown experimentally and beyond any possible doubt that light indeed was a wave phenomenon. So far, so good! Then the question arose: waves of what? Waves are always waves of something—water, air, whatever. You cannot have waves of nothing. It had been known, probably since Galileo's student Torricelli created a vacuum in a glass tube, that light (unlike sound) propagates quite undisturbedly when there is no air present. Consequently light cannot consist of waves of air. The fact that light propagated in seemingly empty space was considered proof that the vacuum itself must contain some other kind of matter which would remain after the air was removed. This led to the conclusion that the ether must indeed fill the entire universe since the light that comes from the very distant stars must be coming as vibrations of the ether.

There were, therefore, very sound reasons for the ready acceptance of an ever-present, invisible ether. It made the process of propagation of light easy to visualize. Waves propagating in a substance were, after all, a commonly observed phenomenon. And the ether seemed to explain such phenomena as electricity, magnetism, gravitation, the propagation of light—none of which were directly perceivable—in a natural and readily acceptable way. As an added bonus, it also made the problem of absolute space more amenable. The mysterious, immovable center of the world was now replaced by the ubiquitous ether. Since ether pervaded space everywhere, it could not possibly move. If, however, it was fixed in space, then it was really at absolute rest, and motion relative to the ether was absolute motion. From this point on, the task of proving the existence of absolute space and determining its framework looked more definite. The only thing that had to be done was to show that something moved relative to the ether!

More and more was learned about the properties of light and by implication about the supposed properties of the ether. Already in the seventeenth century, the Danish astronomer Olof Romer was able to determine in an indirect way the approximate value of the speed of light. It turned out to be enormous. More accurate measurements later showed it to be about 180,000 miles per second.* In other developments, it became clear that the vibration of light waves, in contrast to that of sound waves, was perpendicular to the direction of propagation. Methods were also devised to calculate from measurements the wavelengths of light waves.† These turned out to be incredibly small, much smaller than any objects made visible by a microscope. The curious situation arose then that while, of course, the phenomenon of light itself remained the epitome of something "perceivable," many of light's characteristics could not be directly perceived by the human senses at all. Sensory information about the quantitative, measurable aspects of light, the aspects which interested scientists most, remained always indirect and secondhand. All this was both interesting and important. Nevertheless, in thinking about light, the most exciting question was not how it propagated but the question of its nature. What is light? What is it "made of"? A definite answer to this question, which, one would guess, was as old as human curiosity, was also found in the nineteenth century, and the consequence of this answer was—eventually—the breakup of the classical ideas of time and space.

The answer came not from studies of mechanics but from a seemingly unrelated field: from studies of electric and magnetic phenomena. In the 1860s, James Clark Maxwell, basing his work on the experiments and in part on the ideas of Michael Faraday, formulated the mathematical laws that describe all electric and magnetic phenomena. In describing them, these laws also unified the two, showing them to be two manifestations of the same physi-

* Romer's estimate was about 20 percent less.

† Among the most important researchers of the properties of light were Foucault, Aragon, Fresnel in France and Young in England.

cal reality: the same field. This was the first "unified field theory" and is now called electromagnetism.

Maxwell's equations strongly suggested that light was essentially the wavelike oscillations of the electromagnetic field. A very simplified verbal description of these equations is the following: (1) Temporal changes of the electric field produce a magnetic field which changes in space; (2) temporal changes of the magnetic field create a spatially changing electric field. The equations themselves contained, of course, detailed and precise information as to how exactly these changes were connected in magnitude and in direction, what the role of electric currents and charges was, and so on. There were two particularly interesting consequences of these equations. The first was that from them one could mathematically derive that both electric and magnetic fields propagated in a wavelike fashion in space and time, and that this wave, furthermore, had properties which earlier experiments had shown to be characteristic of light. Last but not least, if one calculated the speed of propagation of this wave, it turned out to be exactly the speed of light. Maxwell also suggested that not only was light itself merely the oscillation of the electromagnetic field, but that there were other oscillations of this same field, and that the reason they were invisible was that their wavelengths were different from those of light, and so they did not excite the human light receptors. Invisible, however, did not mean undetectable. Indeed, in 1888, the experiments of Heinrich Hertz in Germany indeed showed the existence of electromagnetic waves with wavelengths longer than those of light. We now call these (in order of decreasing wavelength) radio waves, radar waves, microwaves, and infrared waves. Another quarter of a century later, the German physicist Max von Laue showed that x-rays are also electromagnetic waves with wavelengths shorter than those of visible light. The radiation from atomic nuclei, called γ rays, have even shorter wavelengths. All these are different manifestations of the same phenomena, all are electromagnetic by nature, and all propagate, in vacuum, at the speed of light.

Like Newton's equations of motion in the eighteenth century, Maxwell's equations of the electromagnetic field were destined in the nineteenth to become the focus of attempts to determine the existence and framework of absolute space. The two sets of laws did not compete with each other because each set described a different aspect of nature. Therefore it was not surprising that they did not relate to absolute space the same way. With Newton's equations of motion, one could not detect inertial motion. This was precisely the content of the Galilean principle of relativity. Maxwell's equations promised a solution to this conundrum. They described, or so everybody thought during the last century, the propagation of waves in the ether. The ether was at rest in absolute space. Consequently if you measured the speed of light while you were standing still in the ether, you would obtain its real value. If, however, you measured the speed of light while moving in the ether (i.e., moving absolutely), you would get a higher or lower value depending on whether you moved against or with the direction of the motion of the light. Moreover, just by showing that this difference existed, a clear proof for the existence of ether and of absolute space would be established. All that remained was to actually perform the right experiments.

The rest, as the saying goes, is history. In the 1880s, the American physicist A. A. Michelson performed several exquisitely designed, accurate experiments with the aim of proving once and for all the reality of the ether at rest in space.* It is also history that while the experiments were accurate enough to detect motion in the ether, no such motion was observed. The existence of inertial motion and absolute space remained as unprovable by experiments using light and electromagnetism as they had by experiments with the motion of bodies in classical mechanics. In hindsight, we know that the concepts of absolute space and time were in fact made quite meaningless by the Michelson experiments. At the time of these experiments, however, nobody realized this, least of all Michelson. He hoped, as did virtually all his contemporaries, that some acceptable reason would soon be found to explain why his equipment had failed to produce the expected results and failed to detect absolute motion in the ether, and that other experiments would still establish detectable differences between motion and rest.

Such hopes were not entirely unreasonable. By the end of the nineteenth century, physicists had largely finished with the exploration of all the then known fundamental processes in nature and thought they had discovered the basic laws of these processes.

* A short description of these experiments is given in the next chapter.

These laws were logically connected with each other by elegant, all-encompassing theories which described nature embedded in the framework of absolute time and space. Virtually all of these laws had been, furthermore, verified by careful experiments and observations. Thus it was felt with some justification that physics was an outstandingly successful mental enterprise which led the way in showing all other sciences how to explore and make sense of the external world. The failure to detect absolute space seemed, in this atmosphere, merely an unpleasant but not really significant blemish. As Michelson himself expressed the consensus of physicists in 1899: "The more important fundamental laws and facts of physical science have all been discovered and these are now so firmly established that the possibility of their ever being supplemented in consequence of new discoveries is exceedingly remote."

But things turned out to be quite different. Physics refused to settle into graceful perfection. It became, instead, the catalyst and agent in one of the most radical transformations in the entire history of human cosmologies.

7

Cracks in the Mammalian Foundation

(The Theories of Relativity)

The universe is real but you can't see it. You have to imagine it.

Alexander Calder

When classical physics broke with the mythological and culturally determined worldviews of all prescientific and traditional societies, this was a radical enough enterprise. Twentieth-century physics, however, turned out to be far more audacious. Not only did it question the validity of earlier human ideas, it challenged the testimony of our senses, the world our brains had created, our inborn models of external reality—the mammalian cosmology itself. In particular, it told us that the very intuitions about space and time that we had acquired through eons of biological evolution could no longer be trusted. And all this happened because of some niggling questions in physics remained unanswered.

The revolutions in physics were triggered by a gradual accumulation of small inconsistencies, by a number of problems which, while seemingly not earthshaking by themselves, still stubbornly resisted solution. By the end of the last century, a number of such unsolved problems had accumulated in physics. Most of them were in one way or another associated with the properties of light. True, Maxwell's electromagnetic theory had clarified a great deal about the nature and behavior of light. But at the beginning of the twentieth century, two important phenomena were still unexplained. These were the creation and absorption of light by matter and the propagation of light.

Since light is emitted and absorbed by atoms, the desire to understand the creation and the absorption of light led logically to the study of processes which took place inside atoms. Such processes can never be observed directly. The human senses—no matter how sophisticated the sensory aids we may use—can never obtain direct information about atomic processes. The inside of the atom is a domain from which all sensory information is necessarily indirect, secondhand. We shall look more closely at this particular problem in the next chapter and we shall see in detail there that the discrepancy between our inborn models and external reality was by far the greatest when we attempted to understand the world of atoms and subatomic particles. Here we shall consider another example of a phenomenon forever closed to our senses: the propagation of light in vacuum. The speed of light in vacuum is so large that human senses can never perceive it directly.

And it so happened in history that the first widely noticed contradiction between the real world on the one hand and our intuitions about time, space, and motion on the other was the result of observations concerning the propagation of light. These observations were to lead to the abandonment of some of the most fundamental fixtures of classical physics, such as the ether, absolute space, and absolute time. They also led to the rejection of another deep-seated intuition: that space and time were two independent and entirely separate dimensions of existence. This latter intuition was common to all earlier human cosmologies whether scientific, philosophical, or mythological, and its abandonment was by itself quite radical.

The new framework of symbolic time and space, in which the separate character of the two was lost, was suggested by Albert Einstein, in 1905, in the special theory of relativity. This new framework denied the objective meaning of almost all purely spatial or purely temporal properties of the external world. Neither space by itself nor time by itself had, in other words, much meaning within this conceptual framework. Instead, the theory of relativity created a new symbolic construction called "spacetime," a conceptual fusion of time and space. Not only did this concept account for some then new observations which could not be rationally explained in the framework of absolute space and time, it also encompassed and made sense of all those observations which had earlier been satisfactorily interpreted by assuming a separate space and time. Consequently, the new concept of spacetime was fairly rapidly accepted in spite of the fact that it had no connection with direct sensory experiences. It soon came to be as indispensable to contemporary physics as the concepts of absolute space and absolute time had been to classical physics.

This chapter is devoted to a description of how time and space were fused into one concept in the special theory of relativity and to an explanation of why this new symbolic framework offered an understanding of reality so much better than the older, classical one.

The theory of relativity is based on two fundamental statements or postulates: (1) All true laws of physics are absolute, and (2) the speed of light is absolute. From these two postulates, one can develop the whole of the special theory of relativity. "Absolute" in this case refers to any feature of the universe which remains the same to an observer whether that observer is moving relative to it or not.* In attempting to answer the obvious question—why a theory about absolutes is called a theory of "relativity"—let us take a look at some examples of nonabsolutes. When we say that the "speed of a car is 40 miles per hour," we are not stating an absolute fact. For if you are sitting in another car, the first car will seem to be going either faster or slower depending on your own speed and direction of travel. If a car sounds a horn, the pitch of the sound is not absolute either. As you move away from the honking car, the pitch becomes deeper, and the faster you move, the deeper the pitch becomes. It is, in fact, not easy to make "absolute" statements in this sense. But it was a commonly accepted idea before Einstein, as we recall from the previous chapter, that space and time and many of their important features were absolute. It was thought, for example, that provided you made careful enough measurements, the length of a body would

* Unless otherwise specified, "motion" in this chapter means motion on a straight line with unchanging speed.

be the same whether the body moved relative to you or not. It was also thought that the duration of a time interval was absolute; i.e., it was the same whether you were moving relative to a clock measuring it or not. However, with the birth of special relativity, all this was seen to be untrue. This was the big surprise in Einstein's theory—even to Einstein himself. The fact that space and time were "relative" seemed so amazing at the time that it was this discovery which gave the theory its name. We know now that this name is not really appropriate. The theory, as is apparent from its basic postulates, looks for absolute and not relative properties. It does not at all say that "everything is relative." It says, rather, that only absolute properties are really important. But the name "relativity" stuck and, however inappropriate, is here to stay.

The postulate "All true laws of physics are absolute" refers to the fact that these laws would be valid for all observers no matter whether the observers moved relative to each other or not. Consequently, no true law of physics could help us decide which of two systems or observers is "moving" and which is "standing." All we could ever say about them is that they move relative to each other. This principle, called Einstein's principle of relativity, is an extension of Galileo's relativity principle. Galileo, as we recall, realized that no mechanical experiment could ever decide whether a system moved relative to "absolute space" or not. Einstein enlarged on this idea and postulated that no experiment of any kind, mechanical or otherwise, could ever detect "absolute motion" and, that therefore, there is no absolute motion at all—with one exception. That exception is *the propagation of light in vacuum.*

It was precisely this innocent-sounding idea, that the speed of light is absolute, which ended up playing such havoc with our inborn and learned models of time and space. For this idea contradicts our basic sensory experiences, our inborn cosmology, and thus "common sense" itself. Our genetic experience had taught us, corroborated consistently by our sensory experiences, to reckon with speeds in the following manner: Suppose you drive a car on a straight highway with a constant speed of, say, 80 miles per hour, and you are overtaken by another car doing 100 miles per hour. Then the other car has a speed of 20 miles per hour *relative to you.* If a third car is coming in the opposite direction, also with a speed of 100 miles per hour, then this car's speed is

180 miles per hour relative to you. In other words, the first car seems to overtake you "slowly" while the second, coming in the opposite direction, quickly disappears—though both are traveling at the same speed of 100 miles per hour on the highway. This is common experience and does not depend on how fast the cars or anything else is moving. We can imagine, instead of cars, rabbits or deer moving at 20 miles per hour or spaceships moving at 80,000 miles per hour. The reasoning would not change.

Light travels very fast—about 18 million kilometers per minute in vacuum.* We may, at least in principle, set up a situation similar to the one described in the preceding paragraph with the variation that the second and third vehicles are replaced by light rays. Following our previous reasoning, we would expect that when light overtook us, we would observe and measure its speed relative to us to be less than the speed of light coming from the opposite direction. But this is not what happens. When we do such an experiment, what we observe is this: Whether the light overtakes us or approaches us head on, whether we move slowly or fast, the speed of light as measured by us (i.e., the speed of light relative to us) *is always the same* as if we were standing still. In other words, even if we assume that we are moving at 99.99 percent of the speed of light (this is a speed an elementary particle, such as an electron or a proton, may easily reach in a large accelerator) and light overtakes us, we know that we will still measure its speed to be 18 million kilometers per minute. And our measurement will also be the same if the light is advancing toward us. The speed of light is always the same no matter how fast or in what direction we move. It is independent of the motion of any observer. If, furthermore, an object which is emitting light is moving—at no matter what speed—the speed of the light emitted will still be 18 million kilometers per minute. Therefore, the speed of light is also independent of the motion of the source of light. In short, the speed of light is absolute.

This unexpected fact about the speed of light was precisely what Michelson and his co-worker C. Morley discovered in their

* "Speed of light" always means the speed of any electromagnetic radiation (e.g., radio waves, light, x-rays) in vacuum. It is believed that gravitation also propagates with the same speed.

experiments in 1887 in Cleveland, Ohio. Michelson and Morley had no artificial vehicles to work with but had at their disposal a natural spaceship: the earth itself. The earth travels in its orbit around the sun at an average speed of about 60,000 miles per hour, which is approximately 17 miles per second. Michelson and Morley sent off light rays in two directions at right angles to each other. The rays were then reflected by mirrors placed at the same distance from the light source, and came back to the place from which both had been sent out. One of the directions coincided with the motion of the earth, the other was perpendicular to it. It is an easy matter to calculate the times needed for each of the two light rays to get back to their common point of origin. The calculations, or their nontechnical explanations, can be found in virtually all technical or popular books on relativity. We will confine ourselves to noting that if the earth and the light rays were indeed moving in ether (i.e., relative to absolute space), then the two light rays would not arrive back at their common point of origin at the same time. The one which traveled in the direction of the motion of the earth would have to arrive later. But no difference in the arrival times was observed in this or later experiments. The instrument was sensitive enough to detect even a small fraction of the expected difference, but no difference was observed. The Michelson-Morley experiment showed simply that the speed of light relative to an observer is the same whether the observer "moves" or "rests." In the following years and decades, several similar or equivalent experiments, some direct and some indirect, have been performed. All have proved the absolute character of the speed of light.

As I mentioned earlier, when Michelson and Morley set out to do their experiments, they had not the slightest notion that the speed of light was absolute. No one had. On the contrary, the aim of their experiment was to determine the speed of the earth in absolute space. The speed of the earth around the sun was, of course, known. But since the sun was also moving inside the galaxy and it was quite conceivable that the galaxy also moved in space, the "real" motion of the earth, its "real" speed in absolute space (or in the ether), was to be measured by this very experiment. The negative outcome of the experiment was actually a blow to Michelson, who had hoped to go down in the history of science as the first person who actually proved the existence of the ether. He became, instead, the first who showed the ether to be a fiction. He was very unhappy about this, but there was no way he could help it. He spent a long time looking for explanations which would somehow save the ether. He repeated the experiments at various times to make sure that the motion of the sun did not accidentally cancel out the effect of the motion of the earth, and refined them in other ways as well. Nothing helped: there was no way to detect the ether; the speed of light showed itself to be absolute.

At first hearing, this description of the behavior of light seems incredible. It is like finding out that "in reality" two apples and three apples make three apples. It completely contradicts our ordinary experience of speed, and this is why we are inclined to say that it contradicts common sense. This contradiction arises from two circumstances: that the speed of light has never been—and cannot be—observed by our senses directly and that the behavior of light had never been relevant either to biological or human cultural evolution. Consequently, we evolved neither a sensory nor a symbolic model of nature to allow for the strange behavior of light. But once we learned about it, we had to make sense of it, and it was Einstein who showed us how.

Einstein's attitude was, in essence, that the fundamental facts of nature have to be taken as they are. We cannot hope to explain right away "why" the world is the way it is. If deeper probing shows that the universe is different from what our immediate sensory experiences indicate, then all we can do is try to use our brains to develop a new framework, a new symbolic model of time and space—in this case, one which is able to take into account the unexpected behavior of the speed of light. What does the speed of light have to do with space and time? "Speed" is a composite idea which is made up exclusively of two simpler concepts: (1) distance and (2) duration. "Speed," or "velocity," means a certain distance covered in a certain time. It has nothing to do with any of the other properties of the moving object. Sixty miles per hour means the same thing whether it refers to a car, to a cheetah, to an electron, or to a galaxy. Consequently if, as in this case, "speed" itself behaves "strangely," as the speed of light certainly does, we should not blame this "strangeness" on some new and perhaps hitherto undiscovered properties of matter or of light (though such unsuccessful attempts were made in prerelativity physics). It must be blamed on our conception of distances, durations, and their relations. In other words, on our ideas of space and time.

Instead of directly attacking the concepts of absolute space and absolute time, Einstein set out in his 1905 paper to investigate what the two postulates of special relativity say about time and space. The first concept he analyzed was *simultaneity,* which is seemingly a fairly commonplace idea. No technical physics paper had ever questioned or dealt with this notion before. Few had seen problems with it. Whenever two events happened, they either happened simultaneously, i.e., at the same time, or they did not. This was true no matter what the events were. There seemed to be little ambiguity in this concept. But Einstein realized that the two axioms of special relativity play havoc with the simple idea of simultaneity. They make it a "relative" idea. This relativity, in turn, demolishes the "time" that had evolved in science and in philosophy during the preceding centuries.

The following paragraphs are devoted to a detailed analysis of Einstein's theory. The reasoning employed here is not abstract; it is not mathematical, and it is not complicated. And it is rewarding to follow it through because the fascinating conclusion contradicts both what we think of as common sense and our everyday notion of "time," and also because by doing so we can see the arresting simplicity of an argument which revolutionized the idea of time more than any other event in known history. The only things to remember throughout are that the speed of light is the same for all observers no matter how they move, and that, according to the principle of relativity, it is not possible to decide which observers are "moving" and which are "at rest."

Question: How will an observer be able to say that two events which occur at two different places are simultaneous? Before attempting to answer, let us agree first on how a person gets information about any distant event. How did I learn, for example, about Neil Armstrong's setting foot on the moon? By watching it on television, of course. At a certain moment, I saw him stepping on the surface of the moon. The two events, the landing itself and my seeing it, were not simultaneous because a certain length of time, about a second and a half, was needed for the electromagnetic wave to travel from the surface of the moon to the earth. So I know that my perception of this event was not exactly simultaneous with the event itself but was delayed by about a second and a half. If, a second and a half before I saw Armstrong's famous step, I dropped my pipe, then I knew in hindsight that Armstrong's step and the dropping of my pipe were simultaneous.

Let us now imagine an observer—say, Jack—standing exactly in the middle of a long boxcar, holding a flashlight in each hand. Suppose that at the same instance, he flashes both lights in opposite directions. Then the light flashes will hit each wall of the boxcar at exactly the same time, and, naturally, Jack will observe that the light hit each opposite wall simultaneously. He would say that the two events were simultaneous.

Imagine now that Jack's boxcar moves, say, to the right, relative to Jill who is standing a certain distance away (see Figure 7-1). Jack's observation doesn't change. His whole system—or boxcar—moves, his distance from the two walls remains the same, and since the speed of light remains constant no matter how the system moves, he still observes the two events to be simultaneous. So far so good. But what does Jill see? The speed of light in the boxcar is the same to Jack, as we said, but the speed of light along the path of the boxcar is also the same to Jill—this being the peculiar property of light. The distance the light she sees travel, however, is not the same since the wall on her left travels in the opposite direction as the light coming from Jack's flashlight. So she will see something quite different; she will see the light on the left wall flashing first. The one on the right, which is traveling away from the flashlight according to her, will flash later. And thus Jill does not see the same two events as Jack to be simultaneous.

Now the question arises—which one of them is right? According to the first postulate of relativity, both are right. There is no way one can say which observer is "standing" and which "moving." They are both equally trustworthy, according to the relativity principle, to describe any event. We must therefore conclude that when we say, "Those two events are simultaneous," we are not uttering a meaningful statement. We must also say to which observer they appear to be simultaneous.

This simple observation plays havoc with the whole idea of time as an objective and unique dimension of our existence. To

Figure 7-1. The thought experiment to examine simultaneity.

appreciate this we have to think a bit more about the meanings of the words "simultaneity" and "relative." The relativity of simultaneity means that we cannot say that two distant events are "really" simultaneous. Their temporal separation depends on the motion of the observer relative to the events. This puts the idea of simultaneity approximately on the same level as the ideas of "left" and "right." One cannot seriously argue whether a house is "really" on the left side of the street or on the right side. What is the left side to me is the right side to somebody standing facing me. This state of affairs may cause problems for a three- or four-year-old, but the child will soon get straightened out. To argue about whether two distant events are "really" simultaneous, whether they "really" happen at the same time, makes about as much sense as to argue about which is the "real" left side of a street. Your answer, in this case, depends on whether and how you are moving relative to the events.

Consider two people. One moves relative to the other in a straight line with uniform speed. Both are, in other words, in inertial systems. Suppose also that these observers have clocks with them which are identical in every respect. You cannot "really" synchronize two clocks at different points in space. If one observer sees the two clocks showing twelve o'clock simultaneously, another equally trustworthy observer will say that they don't show the same time at all, but that one is ahead of the other. The various observers can never ever agree precisely because simultaneity is an observer-dependent, "relative" concept. *But if you can never ever synchronize two clocks at different places in the absolute sense, then you cannot speak about "the time" as being the same at two different places either.* Even less can you speak about it being "the same time" everywhere. If all the clocks in the universe showed four o'clock, say, for one observer, the same clocks would show widely different times to other observers who moved relative to the first. All this has nothing to do with the specific ways our clocks are constructed. What we are facing here is an intrinsic property of time, not of measuring instruments. And as a consequence of this property, Newton's "equably flowing time without relation to anything external" turns out to be an illusion. There are as many "times" as there can be "observers" moving relative to each other.

It should not be surprising that these rational, scientific arguments about impersonal observers and imaginary clocks aroused a furious emotional response when they were first proposed. Time, to start with, had never been too comfortable an idea to think about. It moved inexorably even while one was thinking, and one could not stop it or slow it down or redirect it. It swept one along and gave no security. If it ever offered any comfort at all, that came from its universality, from the notion that it flew uniformly and unchangeably. It was democratic in the sense that death is. Its flow was always the same for everyone everywhere. You could not stop it, but at least you could resign yourself and swim with it, and it would carry you the same way it carried everybody and everything else. The statement of the relativity of simultaneity took away even this comfort. There is, it said, no such thing after all as the same time everywhere and no such thing as time flowing at the same rate everywhere. There is no universal time. What we had discovered was that not only is the river of time unstoppable, but there is, in fact, *no single river at all.* There are only disconnected streams. "Time," thought to be a unique dimension of our existence, turned out to not exist as such.

It sometimes helps to avoid confusion if we remember that the relativity of simultaneity refers only to the simultaneity of distant events. If two events occur at the same place simultaneously, they are simultaneous to all observers. If it refers to the same point in space, simultaneity is absolute. This stands to reason since, if the two events happen at the same place, the absolute character of the speed of light does not come into play and there is no ambiguity in the idea of simultaneity. So when Brutus' dagger met Caesar's heart, Caesar died in every system of reference. Death remains absolute in relativity—as do lovers' kisses, since they are exchanged at the same time and at the same place.

After the first storms, the opposition to the relativity of simultaneity died down surprisingly fast. In a matter of about ten years, the special theory of relativity became a universally accepted part of physics, and the problem of simultaneity no longer seemed to bother physicists. This happened mainly because the theory was so successful in explaining observations and also because its inner logic became gradually apparent to all. From a purely logical standpoint, this fairly quick acceptance of relativity seems natural. Indeed, physics texts emphasize that the early opposition to Einsteinian simultaneity resulted from confused thinking—and physicists therefore are not surprised that the resistance to the new ideas soon petered out.

The view of a psychologist, however, may be different and is probably quite justified. The judgments of human beings are not always determined exclusively by logic—particularly when such an emotional topic as the nature of time is involved. Jean Piaget, for one, thought the initial opposition to the relativity of simultaneity quite natural but found it perplexing at first that this opposition died down so quickly. How was it possible, he asked, that the discovery of the relativity of simultaneity did not cause greater, more persistent, and longer-lasting opposition? Why did it not cause a deep and long-lasting crisis at least in philosophy? To try to explain this, Piaget invoked the results of his own experiments dealing with the perception of simultaneity by young children. According to his results, simultaneity is not a property of time which young children are able to perceive in any precise way.

In one of the Piaget experiments, children were shown two dolls walking in the same direction. As long as the dolls started together, walked side by side at the same speed, and stopped together, the five- and six-year-old children perceived both their start and finish as being simultaneous. But if one doll moved faster, got ahead, yet actually stopped at the same time as the other, the children insisted that the one farther out had stopped later. Simultaneity therefore seems to have a different, more restrictive meaning at an early age than in later years. Older children had, of course, no difficulty in perceiving the simultaneity of the stopping of the two dolls. But these experiments tell us that simultaneity is not a primitive, inborn, or early acquired notion. It is a later acquisition, more of an intellectual construct, which evolves at a more advanced stage of the development of the human brain. And this was the reason, Piaget argues, that it was relatively easy to abandon it and replace it with a new and different intellectual construct: Einsteinian simultaneity.

Another interesting idea originating from experiments of the Piaget school, one which I mentioned earlier, is that the notion of speed is more fundamental in young children's perception than is that of any other perceivable property of time, such as duration or simultaneity. "Speed," in other words, is a more basic, more intuitive, and less derivative notion than are our concepts of the purely temporal features of motion. This is a remarkable result. As the physicist David Bohm points out, it may help us to understand why we were able to construct intellectually without too much difficulty the symbolic space and time framework of relativity around the speed of light. For when we built up the space and time of relativity around a "speed," we may have repeated, on a much more abstract level, what we had done with our first mental constructions of time and space in our childhood.

The thought experiment with rays of light from which the relativity of simultaneity was deduced is a simple example of the reasoning used to examine the properties of time and space in the special theory of relativity. As the theory unfolded, mathematical arguments and idealized experiments followed one another, and more and more features of space and time turned out to be relative and to have little precise scientific meaning. The distance between two fixed points in space, for example, and therefore such a simple thing as the length of a rod, turned out to be relative. "Relative" meant the same thing here as it meant in connection with simultaneity. Suppose I have a rod in my hand which I measure to be one foot long. You whiz by me parallel to the rod at high speed, measure the length of the rod, and find it to be, say, one-half foot long only.* But here comes relativity: If you also have a rod in your hand the length of which you measured to be one foot, I will also measure your rod to be just half a foot long. Both results stand to reason. We must both see the world in the same terms because there is no way to determine which of us is moving. Other observers may find a great many different lengths depending on how fast they move relative to me or relative to you. This is not due to some technical inability to measure the lengths of moving rods accurately. The results of such experiments, like the results of the thought experiments on the relativity of simultaneity, are due to the basic characteristic of time and space—the behavior of the speed of light. This is why the length of a body depends on how fast you move relative to it, and this is why the concept of length or distance is not terribly meaningful. The volume of a body is also relative. Its measure again depends on whether the observer is moving or is at rest relative to it. The duration between two events is not absolute either, and, consequently, the lifetime of any process is also relative. It all depends on how fast the measuring observer moves relative to it.

Among all the relative properties of time and space, it was just this property of time, the *relativity of durations,* which was first confirmed by accurate measurements in the real world. More than forty years ago, in the late 1930s, it was discovered that when cosmic rays (which consist of atomic nuclei, mainly protons, coming from somewhere in outer space and carrying a lot of energy)

* In order for you to arrive at this particular answer, your speed relative to me must be, according to Einstein's formula as evaluated by my pocket calculator, about 87 percent of the speed of light.

hit the atoms of the outer layers of the earth's atmosphere, these collisions produce certain elementary particles, among which are a type of particle we now call the "muon." These particles are unstable: after some time they decay into electrons and other particles. The muons' average lifetime, when they are at rest in a laboratory, was measured to be about two-millionths of a second. "Average lifetime" means approximately that if a hundred muons were created at some time, then two-millionths of a second later, there would only be about fifty muons present and after four-millionths of a second, only twenty-five. A few hundred-millionths of a second later there would be virtually none left. The muons created during the collisions in the atmosphere travel very fast; their speed is quite close to the speed of light. But even if the muons traveled exactly at the speed of light, they could not cover in a few millionths of a second more than a couple of kilometers. The outer layers of the atmosphere, on the other hand, are about 10 kilometers from the sea level. How then was it possible that these muons were found in large numbers at sea level? The answer is that when they travel that fast relative to an observer at rest, their time flows at a slower rate. To an observer who is at rest on earth, the muon lifetime would look longer than the one measured when the muon was at rest in the laboratory. It can be calculated that if the muon moves with a speed slightly higher than 99 percent of the speed of light, then its clock slows down by a factor of 10; this is sufficient to explain how it is able to reach sea level.

This is a convincing explanation, but now one can ask the following: If it is true that all observers are equally trustworthy, then how would a hypothetical observer *who moves with the muon* explain all this? This observer would not see the muon moving at all. In fact, he or she, at rest relative to the muon, would measure the lifetime of the muon to be just two-millionths of a second. How would our observer explain how the muon manages to travel through the atmosphere in such a short time? The answer is that this observer would not measure the distance to be covered as 10 kilometers. Recall the example of the rod and the shrinking of distances to moving observers? The observer sitting on the muon would find the atmosphere moving very fast, would consequently find the distance through the atmosphere to be shorter, and thus would have no difficulty in understanding how the muon covers the shorter distance in the shorter time. Neither the observer at sea level nor the one sitting on the muon should see any paradox in this as long as they both use the relativistic concepts of space and time. Both would find the arrival of the muon at sea level impossible to explain if they thought of it in terms of classical Newtonian space and time.

And what about the propagation of light? With the disappearance of the ether, what was the medium of which light waves were formed? If there was no ether, then what were light waves waves of? The answer of relativity is that light waves (and electromagnetic waves in general) are autonomous oscillations. They propagate by themselves in completely empty space. It is their electric and magnetic properties which oscillate and spread in a wavelike fashion.

This is a remarkable answer for two reasons. First because the idea of autonomous oscillations which exist without a carrying medium, while not particularly strange to our ears, was quite unbelievable to many of Einstein's older contemporaries. "We know that light is a vibration of something—of what?—that 'what' we call ether," said Michelson, the very individual whose experiments constituted one of the best pieces of evidence for relativity. It is always difficult to break decisively with ideas we learned in our youth. Einstein's older and more traditional contemporaries were attached to the concept of the ether to an extent which we might consider sometimes irrational. The main reason why the concept of the ether was so attractive might well have been that the propagation of light in a medium was a process which fitted the internal models of our brain very well. It was congruent with our experiences. The autonomous oscillations of electromagnetic propagation are less easy to visualize and more difficult to fit into our mental models of the external world.

The speed of light is the only absolute speed in nature, but one can find other absolutes as well. In looking for those, it is helpful to remember that neither purely spatial nor purely temporal properties (other than one important temporal absolute which I shall describe presently) can be absolute, but only some combination of both. The speed of light itself is such a combination, but there are others. An interesting and typical absolute measure is the so-called "event-distance." Imagine, for example, two events which occurred at two different places at two different times. (It does not matter at all what events one imagines. My touching the paper with my pen as I started writing this sentence can be one event, and the sight of an explosion of a supernova in a distant galaxy another.) Measure the spatial distance between the two places and call it a. Next measure the time elapsed between the two events. Now calculate the distance light would have covered during the time interval you just measured. Call this distance b. We know that neither a nor b has much objective meaning since neither of them is absolute. Observers moving relative to each other could find widely different values for a as well as for b. But if they take the difference between the squares of the two (i.e., $b^2 - a^2$), this number will be the same for all observers (assuming they use the same units). While spatial and temporal distances are in themselves relative and devoid of much significance, the event-distance as defined above is an absolute quantity, an objective property of the external world. It does not depend on how an observer moves relative to the events. But this absolute property of nature is something we have no perceptual model of. We have no senses to perceive and no intuitive model to imagine event-distances. They are parts of the scientific model of reality; they belong to a conceptual cosmology whose mental picture of the world is framed by a unified spacetime.

But as I just mentioned, there does exist one purely temporal property of nature for which the temporal order is absolute and which fits our mental models very well. The temporal order of two events can never show itself reversed for any observer if one of the events could be the cause of the other. Suppose A shoots B from a distance. The time interval between A pulling the trigger and B's death can seem and can be measured to be different by different observers since different observers may find the speed of the bullet to be different. But no observer can ever see B dying *before* A pulls the trigger. It is crucial that this be so. If all observers have equal standing in observing the real world, then the world must be equally rational to all of them. In a rational world, "cause" always precedes "effect." The theory of relativity assures this by demanding that no form of matter, no form of energy can ever propagate, no information can ever be transmitted faster than the speed of light. Once this is assumed, the mathematical formalism of the theory guarantees that if one observer sees event C causing event D, then the temporal order of the two will remain the same for all observers. The speed of light, therefore, is not only an absolute speed in relativity, it is also a *limiting* speed. No information, no signal can ever move faster. Since the movement of particles from one place to another can always constitute a signal or information, therefore no particle can ever travel faster than the speed of light.*

The very existence of a limiting speed makes it possible to consider two different events which cannot be connected causally. Imagine two persons, A and B, positioned so far from each other that light needs five minutes to travel from A to B. A is firing a gun when an observer's clock shows noon. When the clock shows 12:02 P.M., B drops dead. Since the speed of light is the limiting speed, the firing of the gun cannot possibly have caused the death of B since the bullet would have had to travel two and a half times faster than light to reach B in two minutes. The fact that B died after A fired the gun must have been purely coincidental in this case. No one, not a Sherlock Holmes, nor even a Nero Wolfe, could possibly fix the blame on A.

In such cases the temporal order of the two events is no longer of importance. Another observer moving relative to our first observer could see B dying before A fires. Since the two events were unconnected, this observer would have seen nothing irrational. The theory therefore can allow this to happen. The temporal order of two events is absolute only if there could be a causal connection between them. It is irrelevant otherwise.

The recognition of all this amounts to finding an interesting new feature of spacetime. In classical space and time, the temporal order of two events was always absolute. If event C happened before event D for one observer, it happened so for all observers.

* It is, however, quite all right to have purely geometrical velocities having any speed we like. Imagine, for example, that a spot is produced on the moon by a beam of light originating from the earth. Even a slow turning of the flashlight would make the spot move faster than the speed of light because the moon is quite far from us. But in such a case what moves is merely a geometrical point—the point where the photons from the beam hit the surface of the moon. This "motion" does not involve any material object or any energy, and no information can be transmitted by it. Nothing physical moves faster than the speed of light in this process, neither the photons nor the torch.

In classical space and time therefore, C was always in the past relative to D (or, what amounts to the same thing, D was always in the future relative to C). Consequently, any two events in classical space and time either happened at the same time or one was in the past relative to the other absolutely—i.e., for all observers. In relativity, there are more absolute possibilities. One event can coincide in time with the other for all observers if the two events occur at the same place. Second, one event can be in the past relative to another for all observers if the first could be the cause of the second. Finally, as in our gun-firing example, one event can be neither in the past nor in the future relative to another, and also not contemporaneous with it—it can be in the "absolute elsewhere." The word "elsewhere" means that the temporal order of the two events—if they cannot be connected causally—can be different for different observers and that, therefore, it is meaningless to label one as being earlier or later than the other.

There are no limiting speeds of any kind in classical space and time. You can add speed upon speed as long as you want. Imagine yourself sitting on a bench beside a railway track. A long train is going by at, say, 99 percent of the speed of light. Imagine further that on this long train there are tracks on which another train moves in the same direction whose speed is again 99 percent of the speed of light *relative to the long train.* Then according to classical reasoning, you would measure the speed of the second train relative to you as 198 percent of the speed of light. One can further imagine that there are tracks on the second train as well and a third train moves on it, again with 99 percent of the speed of light relative to the second train, and so on and so on. One is then able to observe as high a speed as one wishes. All these assumptions are in accordance with the properties of classical time and space.

But the picture is rather different in reality. The mathematics of relativistic spacetime is different from the mathematics of classical time and space. Consequently, the rules of "addition" of two or more speeds are also different in relativity. If you combine the speeds of all these trains using the rules of relativistic spacetime, you will get an entirely different picture: you will never find the resulting speed to be greater than the speed of light. In fact, the formula for combining speeds is such that if you "add up" the speed of light as many times as you wish, the resulting speed will still be just the speed of light, and not more. The limiting character of this speed is, in other words, built into the mathematical framework of relativity.*

At this point, a legitimate question arises: If the spacetime framework of relativity is that much different from classical space and classical time, and if it is the relativistic framework which better describes reality, then how was it possible for classical physics to spend three centuries of development on such meaningless illusions as absolute space and absolute time? Why did it take three centuries to discover the fictitious character of absolute space and time?

The answer to these questions lies in the fact that the speed of light is very large compared with all other speeds we are likely to encounter as we observe the motion of various bodies on earth or in the solar system. To see this point better, let us imagine that instead of being in our actual universe, we had evolved and lived in another one where the speed of light—i.e., the only absolute speed and also the highest speed attainable by anything at all—was rather small, so that our senses could easily perceive it.† Let

* The hypothesis that the speed of light is an upper limit for all velocities was thought to be correct for about sixty years following the creation of the special theory of relativity. More recent theoretical analysis has shown, however, that the existence of particles moving faster than the speed of light would not contradict the theory of relativity provided these particles *always* moved at speeds higher than that of light. Some people then suggested that such particles may exist. The speed of light, according to their analysis, still remained absolute and it also remained the limiting speed. But for these—as yet wholly hypothetical—particles it was a limiting speed from below. These particles, called "tachyons" (*tachys* is Greek for fast) cannot, in other words, ever move slower than the speed of light. It has also been shown that the existence of such particles would not necessarily lead to violations of the causal order of the world if their observable effects were carefully interpreted. During the seventies, several experiments were set up with the aim of detecting such particles. But the hunt so far has been unsuccessful.

† Never mind that this might not be possible at all. If the speed of light were much different from what it is, the working speed of our nervous system might be correspondingly different since molecular interactions are always electromagnetic. Therefore the speed of light would probably be felt by living things to be very large whatever its actual value was. Nevertheless the above assumptions help to illuminate the relation of spacetime to space and time.

us further assume that in this imaginary universe, the laws of physics are the same as in the real world and that the laws of biological evolution are also essentially the same. In such a universe, the perception of a separate space and a separate time could never have evolved in higher animals because such perceptions would not have been adaptive. There would be a "map" in our brain for spacetime but no separate spatial or temporal maps. The latter would be of no use. We would have no separate words for "space" and "time" either, only the word for "spacetime." We would live in a world where every smart six-year-old would understand the relative character of lengths and time intervals just as easily as the children in our world understand the relative character of left and right. In this imaginary universe, all information would be processed in the brain through models of the world in spacetime. The human equivalents in such a universe would have to do some hard thinking to realize that if their limiting speed were to be much larger, then the world could be perceived in separate space and time. In contrast to our actual universe, in other words, where we have evolved intuitions for time and for space separately while spacetime is an intellectual construction, in our imaginary universe we would be more likely to have evolved an intuition for spacetime while separate space and time would have evolved as hard to grasp abstract concepts.

We can now imagine another universe where there is no absolute and no limiting speed at all. This means that some signals (say light) could travel with infinite speed. A strong enough flashlight would illuminate all parts of the universe the moment it was switched on. In such a universe there could be no theory of relativity (the second axiom would become meaningless), and space and time would truly be both absolute and independent of each other. "Spacetime" would be unthinkable.

In our actual universe the speed of light is not infinite, but it is very large, so large that we can never perceive it. Therefore, as far as our senses are concerned, it may as well be infinite. Consequently, our nervous system evolved as if the speed of light were indeed infinite and therefore it perceives the world in separate space and time. It was only much later, through our ability to devise ways to measure the speed of light and to ascertain its absolute character and through our ability to reason abstractly, that we could come to the fascinating conclusion that separate spatial properties as well as separate temporal properties are, in fact, largely illusions. They do not help us make sense of our more demanding observations of the world. "Spacetime," in contrast, helps us do so and thus it becomes our "real world."

This scientific "real world," however, is not identical to the everyday world, or even to the everyday scientific world. If we have to walk 10 miles, the effort required does not become illusory because we know that for some fast-moving observer the distance may be just a few yards. When we say that the Andromeda galaxy is 2 million light years away from us, we do not point out that this is true only in a system where both we and the galaxy are at rest. The illusion of separate time and space is and will always remain very important to us. But now that we know the limits of this illusion, we know how far it can take us—and we know when to leave it behind.

The importance of relativity is twofold. Apart from correcting and clarifying our perceptions of the world and allowing us to understand the nature of space and time better, it is also vitally important as a practical research tool. While our senses can never perceive speeds close to the speed of light, elementary particles as well as distant galaxies do move relative to each other at speeds quite close to that. Therefore, without using the concepts of relativistic time and space, we would be unable to make sense of what we observe in those realms outside the readily perceivable: the worlds of the very small, the very big, and the very fast. Like all adaptive models in history, "spacetime" enlarges the understandable world.

The fundamental laws of physics are absolute, says the first postulate of the special theory of relativity. But a great many important laws of physics were discovered before relativity was born. Are they then all wrong? If some are, how do we decide which ones are and which ones are not? We recall that, at the end of the nineteenth century, classical physics consisted essentially of two great theories: classical mechanics and classical electromagnetism.* It is a mathematical fact of life that both these theories cannot be absolute. If one of them is, then the other is certainly not. The Michelson-Morley experiment was essentially a vote for

* I ignore thermodynamics as it is not relevant to our subject.

Maxwell's theory and, consequently, a vote against Newton's equations of mechanics. The theory of relativity assumes, in other words, that Maxwell's equations of electromagnetism are a prototype of a true law of nature. Relativity does not change these equations at all.*

But if Maxwell's equations are a prototype of the true laws of physics, then Newton's mechanics, the Newtonian equations of motion, are not. In order for us to have absolute laws in mechanics as well, Newton's equations had to be changed. The theory of relativity gave a clear prescription for these changes and they were accomplished without much difficulty. In the process a new branch of physics, what we call relativistic mechanics, was created. It was in the context of relativistic mechanics that many of those surprising new facts arose which contributed to the fame of the theory. The new mechanics predicted, for example, that the inertial mass of a moving body would increase as the speed of the body increased. More precisely, it said that if a body moved relative to an observer, that observer would measure its mass to be larger than another observer who was at rest relative to the same body. The increase of mass is such that it is a practical impossibility for any body ever to reach the speed of light. This gives a physical underpinning to the theoretical necessity that no signal can travel faster than the speed of light. Another famous discovery concerned the two important conservation laws in classical mechanics, the laws of the conservation of momentum and the conservation of energy. Neither law could be given an absolute form. Einstein, however, was able to create a new and absolute law of conservation by redefining and unifying the ideas of energy and momentum. The unification of energy and momentum was achieved by the same mathematical method as the unification of time and space. From this new, unified, conservation law emerged yet another law, which connected the conservation of energy with the conservation of mass: $E = mc^2$. This equation was destined to become the most famous scientific law in history.

The unusual fame of $E = mc^2$ is due to its perceived connection with nuclear energy. This connection is, however, often mis-

* A glance back to page 142 will convince you that these laws display a symmetry in that space and time play similar and interchangeable roles. These laws describe changes in space and time in an absolute manner.

understood. The equation itself makes no reference at all to nuclear energy. When Einstein discovered this law, he could not possibly have known that atoms have nuclei at all. That idea was only suggested by Ernest Rutherford after a series of celebrated experiments during the years 1908–1911. The connection between nuclear energy and the $E = mc^2$ equation became established only in the early 1920s when it turned out that Einstein's law could be used to measure the energy contents of atomic nuclei. The equation says that the energy of anything can equally well be considered as part of its inertial mass. Therefore, mass can be thought of as energy. And the equation gives the prescription for calculating the energy content of a certain amount of mass (or the mass of a certain quantity of energy). The factor which converts one to the other happens to be the square of the speed of light. The equation expresses a general law valid for all natural processes involving masses. It is just as valid for the burning of a candle as for the firing of a nuclear reactor. In the case of a candle, however, the energies involved are too small to give a measurable change in the mass, and, therefore, the relation between energy and mass does not manifest itself in an observable way. But the energies which nature put into atomic nuclei are huge, and their mass equivalents manifest themselves in a measurable way.

The reason why physicists took such an immediate interest in this equation was that, throughout history, mass and energy had been considered two distinct and independent characteristics of matter. The $E = mc^2$ equation shows that the two are actually equivalent. The main significance of this result was that it simplified the world and made it more coherent.

The rewriting of Newton's laws of mechanics was extremely successful. It led to the discovery of new laws and new phenomena. But the other great product of Newton's genius, the universal law of gravitation, still remained an enigma. The law itself was not absolute since first of all it said that the gravitational force between two bodies depends on the spatial distance between them. That statement alone would have made the law suspicious since distances are not absolute. A more exact analysis showed that the Newtonian law of gravitation was indeed not absolute because observers moving at different speeds would find the law taking different forms. For some years, Einstein and others tried to change the law of gravitation to make it absolute, to make it conform to the spacetime of special relativity. All the attempts were failures. But from these failures was born, ten years after the special theory, a theory which once again changed the concepts of space and time and did so in yet other unexpected ways. This was the general theory of relativity.

It was around 1907 that Einstein became convinced that the problem of gravitation could not be separated from the problem of acceleration.* The idea that connects the two can be easily understood through a short piece of science fiction. Suppose that, like Rip van Winkle in Washington Irving's story, you fell asleep, and have woken up many years later in a closed room. You do not know where you are, but a note on the table tells you that you have been kidnapped and deposited somewhere. Two alternatives are given as to your whereabouts with the understanding that one of them is true. Either you are in an ordinary house somewhere close to where you normally live, or you are many millions of miles away from the solar system in outer space in a rocket-equipped spaceship. The note asks you to decide by means of physical experiments which of the two alternatives corresponds to reality. If your guess is correct, you will be rewarded by a speedy transport home. If it isn't, some terrible fate will await you.

At first your problem seems quite simple. You remember all those astronauts you saw on television floating around weightlessly in spaceships and notice immediately that you do not float at all. Your feet are planted firmly on the ground, as is the furniture in the room. Then you notice a ball somewhere in the room and decide to use it to actually test the law of free-fall. When you do so, you find that the ball falls about 5 meters in the first second,

* Acceleration is the quantity which measures how rapidly the velocity of a body is changing. A freely falling body on earth, for example, increases its velocity by about 10 meters per second during every second. After falling for one second, its speed is 10 meters per second; after falling two seconds, its speed is 20 meters per second, and so on. Free-fall incidentally, is an example of what is called "uniform" acceleration where the rate of increase of speed remains the same throughout.

about 15 in the second, 25 in the third, and so on.* You rediscover, in other words, the law Galileo discovered almost four hundred years ago. All these facts seem to establish that you are nowhere near outer space but resting safely on good old earth.

But can you really be sure? In fact, even after all these experiments, the answer is still in doubt. It is still an even bet that you are in a spaceship far away in outer space. For what if the rockets attached to the spaceship are firing smoothly and noiselessly in a direction that is "downward" to you, and are thus increasing the "upward" speed of the spaceship by 10 meters per second. If that is the case, your feet and everything else in the room should still be pressed against the floor just as they would be on earth, and the ball would seem to fall exactly as it would on earth. And you would have no way of knowing that the cause of all this is not gravitation—there is no large mass anywhere near you—but actually *inertia,* which is the same force you feel whenever you are in a vehicle which is changing its velocity.

You can rack your brains and think of other experiments which would distinguish between the effects of gravitation on earth and of uniformly smooth acceleration in an enclosed system. But nothing will work. There is no way; no instrument, and no calculation, could answer the question: Are you on earth or are you in a uniformly accelerating spaceship? In other words, are you attracted by a mass or are you merely accelerating? For you in your room, the effects produced by these entirely different external circumstances are going to be exactly the same.

Making a bold generalization, Einstein concluded that the effects of gravitation and acceleration must be identical. And we cannot distinguish between them not because we are clumsy or stupid, but because there is actually no difference between them. In any small region of space, gravitation and acceleration are one and the same phenomenon.

Einstein made this assumption of the equivalence of gravitation and acceleration, commonly known as the "equivalence principle," the starting point of his general theory of relativity. And while this principle in some forms had been known since the

* This data corresponds exactly to the acceleration data described in the previous footnote. If they do not seem to agree, remember that the speed is changing constantly during free-fall, even during fractions of seconds.

beginnings of modern science—Galileo's discovery that all bodies fall freely in the same way was already a statement of this principle—it was not until Einstein analyzed it that certain important consequences of the equivalence principle were uncovered.

What were these consequences? In order to find out, let us decide that we shall try to make sense of every observed phenomenon by assuming that the principle of equivalence is a basic law of nature, and let us stick to this assumption even if it temporarily seems to contradict common sense. Now imagine again that you are in your unidentified room, but that this time you discover a window in it and can see that you are actually in a spaceship far removed from any star, planet, or other concentration of mass. If the ship accelerates, you feel the effects, which, as we just stated, are the same as the gravitational effects of a nearby mass, but now you know that the effects do not come from any external body but are the result of your own inertia. Since we have taken the principle of equivalence as a basic law, we are bound to say that if the effects of acceleration are not due to a force exerted by an external body, then the identical effects of gravitation are not likely to be caused by forces originating from an external body either. The conclusion we are bound to draw therefore is that there can be no such thing as a "gravitational force" emanating from a mass.

While this may sound quite absurd in light of everyday experience, we shall nevertheless stick to it. And to see where it leads, let us concentrate on accelerating systems and try to learn from them about the nature of gravitation. Suppose you are in a system which accelerates upwards in such a way that its speed increased by 10 meters per second every second. Imagine then that you are standing at a wall and throwing a heavy ball across the room. Since the system accelerates upwards while the ball is flying, the ball will hit the opposite wall at a point lower than the one it started from. The ball will behave exactly, in other words, as it would in a normal room on earth; it will fall downwards. No surprise there—this is precisely what the principle of equivalence says: Acceleration is completely equivalent to gravitation.

Next, let us suppose that the dimensions of the ship are so large that it takes light a full second to cross it. You flash a short light impulse across the ship. If the system accelerates upwards uniformly during the time the light is crossing the room, the light should also hit the wall at a point lower than the one at which it started. We cannot actually perform this experiment—we cannot have rooms 300,000 kilometers long on earth, as the whole length of the equator is only about 40,000 kilometers—but we can still imagine it.

Now, both the ball and the ray of light were, as we said, completely free objects.* There was literally nothing, no body nearby, which could have influenced these objects. It would seem, therefore, that in your reference system, Newton's first law of motion was not valid. Such completely free bodies as the ball or the light ray did not move on a straight line with constant speed but behaved as if there were large bodies nearby exerting gravitational force.†

The general theory of relativity explained all these apparent paradoxes in a perfectly logical, straightforward way. The reasoning made use of two assumptions. One was the principle of equivalence, which, as we just saw, implied that there was no such thing as "gravitational force." The second, which is an extension of the first postulate of the special theory, says that the fundamental laws of nature must be valid not only in reference systems which move *uniformly* relative to each other, but also in *any reference system* no matter how complicated or irregular its relative motion may be. The general theory of relativity is based therefore on the strongest possible demands about the absolute character of the laws of nature. There was, therefore, even less justification for terming this a theory of "relativity" than there had been with the special theory.

Let us take a look at the second assumption. It may sound strange at first because, as we saw earlier, Newton's first law of motion seemed to be invalid in an accelerated reference system.

* Light may count as an "object" or a "body" because it carries energy and we know from special relativity that energy is equivalent to mass.

† It is of historic interest to note that all these experiments were purely imaginary. The first experiments in accelerated systems came with the advent of the space age about half a century after Einstein recognized the importance of the principle of equivalence. But it was these never directly performed experiments (called thought experiments) which gave Einstein the clue (around 1912) as to the real nature of gravitation. Once he had found it, it took him three more years to find the mathematical tools to formulate the correct mathematical laws of the complete theory.

Completely free bodies did not move on a straight line at all, they moved on a curved trajectory. Is then the first law of motion, the law of inertia, false? Not at all, said Einstein, if we reformulate Newton's first law in the following way. Instead of saying that a free body would always move on a straight line with uniform speed, let us say that a free body would always move on the shortest path between two points. At first this might seem to be just splitting hairs. The shortest route between two points *is* the straight line connecting them. When we say that, however, we implicitly assume that the rules of geometry in space are the same as those we learned in high school—the rules of Euclidian geometry. But if we give up this assumption, we can still stick to Newton's first law of motion even in an accelerated system. The ball we threw, the light ray we sent across, did move on the shortest route in an accelerated system if the geometry of that system was not Euclidian. If we recall the principle of equivalence, however, we must also say that geometry in the presence of a gravitational field cannot be Euclidian either—and here we discover the essential feature of the general theory of relativity.

If we sum this up in a more coherent fashion, the following picture emerges. Gravitation is not a force at all. Gravitational interaction does not exist. But if you place a body somewhere in space, then *the mass of that body influences the structure of space around it.* And since we know that we cannot express absolute laws in terms of space alone, we had better correct ourselves right away and say that *masses influence the structure of spacetime.* It is easy to visualize how such a thing can happen by invoking a simple analogy. Imagine a stretched rubber sheet in a frame. As long as there is nothing on the sheet it is flat. If you place a ball that is small and light on the sheet, the ball will probably stay where you put it or, in the absence of any interference such as friction, roll on a straight line with constant speed. However, if you put a heavy ball on the sheet, the ball will depress the sheet, and therefore the sheet will no longer be flat, but curved. And if you then put a light ball on a sheet already depressed by a heavy ball, the light ball will move toward the heavy one, following the depression in the sheet. From the outside, if the sheet were completely transparent, it would seem as if the heavy ball had actually attracted the light one.

This is easy to imagine or to draw. We can, however, neither imagine nor draw the curving of three-dimensional space, let alone the curving of four-dimensional spacetime. But the fact that we cannot construct a sensory model does not mean that we are mentally powerless to deal with such concepts as curved space. In fact, the conceptual exploration of such spaces—in the form of non-Euclidian geometry—preceded the general theory of relativity by several decades.

The first non-Euclidian geometry was constructed as early as the 1820s. It was discovered simultaneously and independently by J. Bolyai in Hungary, N. I. Lobachevsky in Russia, and C. F. Gauss in Germany. Gauss never published his results, probably because he was afraid of the controversy they might have aroused, and, perhaps for this reason, the first non-Euclidian geometry is known today as the Bolyai-Lobachevsky geometry. This is a geometry where all of Euclid's axioms are supposed to be valid except the fifth, which, as we recall from Chapter 4, is the axiom concerning parallel lines. Bolyai and Lobachevsky showed that a logically consistent, perfect geometry can be constructed on an assumption which flatly contradicts the fifth axiom.

This was an important discovery. Throughout its evolution, modern mathematics had often constructed concepts which had no observable models or counterparts in anything real, yet were highly relevant and useful, such as imaginary numbers. Physics could not possibly have evolved as fast as it did if imaginary numbers had not been thought of. But it was the geometry of Bolyai and Lobachevsky that was the first *geometry*—a discipline supposed to be dealing with perceivable lines, curves, and so on—which deliberately contradicted the sensory properties of the world and based its laws exclusively on logic.

In 1854, the concepts of non-Euclidian geometry were greatly improved and generalized by the German mathematician Bernhard Riemann. The mathematical apparatus of the general theory of relativity is based largely on Riemann's ideas, on the ideas of what is now called Riemannian geometry.

The geometry of Euclid—the one which virtually monopolized all thinking about space until the last century—implicitly assumed that three-dimensional space is "flat." It allowed for curved lines and surfaces, but these were always embedded in a flat space in which the shortest route between two points was always a straight line. Riemannian geometry was more general. It said that the shortest route between two points was determined by the structure of space as much as the shortest route between two points on a surface was determined by the shape of the surface.

In the general theory of relativity, the structure of spacetime is determined by the way masses are distributed in it. How it is determined is the content of the fundamental equations of general relativity, Einstein's field equations. These equations were constructed in such a manner that if the distribution of masses in spacetime is known, we should be able to calculate, at least in principle, the geometry of spacetime. These equations connect the geometry of the real world with the way objects are distributed in it. In trying to find out how these ideas work, Einstein applied them first to the solar system. This was the most immediate and logical thing to do. The mass of the sun was known and it was also known that this mass was very nearly spherical. Hence the mass distribution of the solar system was known.* The field equations then allowed Einstein to calculate the geometry of spacetime in the solar system.

All that remained was to remember the basic idea of general relativity: There is no gravitational attraction between the sun and the planets. The only role of the sun is to change the structure of spacetime. In this spacetime structure, the planets experience no force at all. They move like free bodies. If the structure of space were Euclidian or "flat," all free bodies, including the planets, would move on straight lines. But because of the presence of the sun, the free motion (which would take the shortest route between two points) does not go on a straight line, but on a more general curve.† When Einstein calculated the curving paths on which the planets move freely, the results were the same ellipses Newton had calculated three centuries earlier when he had assumed the presence of gravitational forces.

Thus the idea of curved space explains that the planets move on ellipses and in this respect yields the same result as Newton's

* The masses of the planets are much smaller than the mass of the sun and their influence on the geometry of the solar system is therefore negligibly small.
† In a textbook on general relativity, C. Misner, K. Thorne, and J. Wheeler summarize all this neatly: "Space tells matter how to move," "Matter tells space how to curve."

theory. The two theories look equally good so far. But Einstein soon discovered that his theory was better. Not only did it account for the same phenomena as Newton's had, but also for a hitherto unexplained occurrence. It had been observed since the early nineteenth century that the point where the planet Mercury comes closest to the sun in its wanderings is itself moving.* A great many efforts were made during the nineteenth century to account for the magnitude of this motion within the framework of the Newtonian gravitational theory. All these efforts were unsuccessful. Einstein's theory gave the correct result for this motion as well.

This is a good example of how a theory starts to get accepted in physics. It has first of all to recover all the established results of the earlier theory it aims to replace, and then it has to produce improvements on the old one. Both aspects are important. It would be a waste of time to work out a theory which yielded the right result for a small effect like the perihelion advance of Mercury if it could not explain, for example, the much more significant laws of Kepler.

More proof was on the way regarding the validity of the general theory. During a famous 1919 expedition of the Royal Astronomical Society, Arthur Eddington found that the path of a light ray indeed became curved close to the sun just to the extent demanded by the principle of equivalence. Another consequence of the same principle, that time passes more slowly in the presence of a gravitating body, was also verified very accurately. As of today, all the observed facts about gravitation and the structure of spacetime are in agreement with the predictions of the general theory of relativity.

But the great significance of this theory is independent of the question whether it is the final word in the quest for a theory of space, time, and gravitation. If the history of physics is a reliable guide, then there are no—and never will be any—final words on fundamental issues. The general theory of relativity, however, remains the very first successful attempt to unite the main ele-

* The phenomenon is called the "advance of the perihelion of Mercury." Actually the orbit of every planet behaves like this, but only in the case of the planet closest to the sun (i.e., Mercury) is the effect large enough to be observable.

ments of the entire physical cosmology—space, time, matter, and motion—into one supremely logical and beautiful conceptual framework.

The general theory of relativity only becomes a really practical research tool when the structure of spacetime strongly deviates from flatness. When this is not the case, i.e., when spacetime is more or less flat, the Newtonian theory of gravitation and (when the speeds involved are very high) the special theory of relativity are good enough for all practical purposes.* In the real world there are two important cases in which space and time deviate significantly from simple Euclidian flatness. One is when large, compact objects are nearby—when, for example, we are close to a large, dense star. The second is when the distances involved are large—when, for example, we speak about the whole universe. In both cases, space and time behave differently from what we are accustomed to.

To see another important property of the geometry of the universe, imagine that we want to measure the distance between two objects. We would like to be very precise and, in order to eliminate all possible influences from other objects, we will take our objects to outer spacc, far removed from everything else. We let the two objects float freely and far enough from each other that one cannot appreciably influence the other. We also make sure that the two objects do not move relative to each other. Then we measure the distance between the two by sending light signals from one to another. The time it takes for light to travel between them is simply proportional to the distance. Now let us say that we wait for a while and repeat the measurement. What we would discover is that a *later* measurement always shows a *larger* distance. This means that the distance between two apparently "resting" bodies is always increasing in time. This is the phenomenon which is called "the expansion of the universe" and it is a property of *space itself* and not of the two bodies. For the distance increases no matter what the two bodies are. The process itself is very real but

* Newtonian gravitation and classical mechanics were, and in the foreseeable future will continue to be, quite adequate in the planning and execution of space flights.

it becomes measurable at large distances only—distances on the order of those between galaxies.

The discovery that space actually expands was yet another proof of how inadequate the ideas of classical space and time were. As far as the real space, time, and geometry of the universe go, even the concepts of special relativity had become inadequate. If two bodies are at rest relative to each other, special relativity asserts that the distance between the two will be different for observers moving relative to the two bodies at different speeds. While this remains true, we can now see that each observer will also find the distance increasing in time. The observational proof for this effect, i.e., the actual discovery that the universe was expanding, came from the American astronomer Edwin Hubble during the years 1924–1928. Hubble also formulated the basic law governing the expansion. The theoretical possibility of such a process, the idea that the distances in the universe may not stay constant, had already been proposed by the Russian theorist Alexander Friedmann, who in 1922, found solutions of Einstein's field equations which described space itself as changing in time. This notion was quite strange at the time, and the significance of Friedmann's solution was not recognized until Hubble's discovery. Physical cosmology is now based to a large extent on the ideas of Friedmann and Hubble.

The presence of the sun disturbs flat spacetime to such an extent that the orbits of freely moving planets in space are ellipses instead of straight lines. If the sun were much denser or had a much larger mass, the distortions in time and space would be even greater. This would be because the more compact and more heavy an object is, the more pronounced is its effect on the structure of spacetime around it. A manifestation of this effect is that the more compact a heavy object is, the more it bends light rays. At a certain concentration, the bending will be so pronounced that the incoming light cannot continue outward at all—it becomes trapped around or inside the object. The object then absorbs all incoming light but never lets any light out. For a spherical mass, this will happen when the radius of the sphere has shrunk to a certain critical length. This critical length is called the "Schwarzschild radius," and it is proportional to the mass of the object. In the case of the sun, this radius happens to be about 3 kilometers; for a star twice as heavy as the sun, the Schwarzschild radius would be around 6 kilometers.

Suppose now that a spherical star loses its balance and starts to shrink, to collapse.* When the surface of the collapsing star becomes a Schwarzschild surface (i.e., the surface of the sphere whose radius is the Schwarzschild radius), we say that *a black hole has been created.*†

The surface of a black hole separates the inside of this object from the rest of the universe in important ways. If the curvature of spacetime is so strong at this surface that light cannot move outward, then nothing else can either. For saying that something could do so would be equivalent to saying that that thing could move faster than the speed of light. The surface of a black hole is therefore a one-way membrane: Things can go in but cannot come out.

The laws of physics also tell us that the formation of the surface does not stop the process of collapse. While we can never see the last part of the process, since neither light nor any other form of information can ever reach us from the inside of the hole, we can follow what happens through calculations. There is nothing strange about this proposition. In principle we have as much reason to trust such calculations as we have reason to trust those which predict when Halley's comet will return. According to our calculations, the whole amount of matter, the mass of several suns or even masses much larger, will disappear very fast *into a single*

* The general theory of relativity predicts that stars which are significantly heavier than the sun will inevitably undergo such a collapse at the end of their "lives"—i.e., when they have exhausted all energy sources which, until that time, allowed them to resist the inward pull of gravitation. General relativity also tells us that in the case of large stars, the process of collapse cannot be stopped or controlled. The word "cannot" does not refer to practical limitations of any kind. It expresses instead a fundamental property of spacetime. That "the collapse cannot be stopped or controlled" is as profound and inexorable a statement as this sentence referring to ordinary time in our environment: "The passing of time cannot be stopped or controlled."

† The term "black hole" was coined by the American physicist John Wheeler. Speculations about stars which are so heavy that the light cannot escape from them had been around for centuries. It was only after Einstein formulated general relativity, however, that the importance of such objects in understanding the properties of spacetime became apparent.

point. This point is called a *singularity*. What happens at this point we do not know. Important physical properties like the density of matter or temperature all become infinitely large and so does the curvature of spacetime. Both matter and spacetime shrink to a mathematical point. It seems that the profound interrelation between spacetime on the one hand and matter on the other leads to a situation where they simply squeeze each other out of existence. As a result, we face a situation where neither matter nor space and time can exist in the normal sense of these words. Without referring to space and time, however, our brain cannot work. Not only does our visual imagination stop, not only do our words and grammar fail, but even mathematics breaks down completely. Controlled, conscious thought simply stops at this point. All we are able to do is to give the situation a name, a symbol: singularity. This word is nothing by a symbol for the most bizarre symbolic time and space ever encountered. If it signifies anything, then it signifies that our ability to make sense of the world in space and time has reached a limit.

No wonder then that the problems of black holes and singularities have intrigued many people active in research. In the 1960s and 1970s, a great deal of effort went into attempts to clarify such problems. In 1965, Roger Penrose, then of the University of London, and Stephen Hawking of Cambridge University were able to demonstrate mathematically that, at least as long as the laws of the microworld were not taken into account, the nature of singularities was such that the actual universe in which we live is likely to end its existence in just such an unfathomable singularity. This very result inspired further speculation that perhaps the universe not only will end in, but also originated from such a singularity.

It was also from ideas about black holes and singularities that a new and important understanding developed during the last couple of decades. This was the recognition that the beginning of the universe and the beginning of space and time cannot be understood without taking into account the laws which govern the world of elementary particles: the laws of quantum mechanics. The first result which showed that these laws could profoundly change the very nature of such objects as black holes came from Stephen Hawking. He was able to demonstrate that the laws of quantum mechanics can actually make black holes somewhat less than "perfectly black." Some black holes can even "evaporate" through a quantum mechanical process and thus send matter into the rest of the universe. This was an important result, one of the first convincing theoretical demonstrations that the laws of quantum mechanics could play a direct role in the large-scale properties of the universe.

8

Space, Time, and Quanta

(The Return of the Frog)

Nobody knows how it can be like that.
Richard Feynman

Light turned out to be a most extraordinary research tool. As we have seen, studies of the propagation of light led to profound and revolutionary changes in the concepts of time and space. The creation and absorption of light by atoms, however, still remained unexplained, and the study of these problems led to even more radical consequences.

From 1860 on, scientists began to study the properties of light and those of other forms of radiation in a radiation-filled cavity. The walls of this cavity were black and for this reason the radiation inside it was called "blackbody radiation." The object of these experiments was to measure how the energy of the blackbody radiation was distributed over the wavelengths involved and how this energy distribution depended on the temperature of the walls. After all these data had been collected, researchers tried to interpret them in light of the then current theories of physics. All their attempts, however, were unsuccessful. No mathematical expressions could be derived from the then known laws of physics describing the expected distribution and temperature dependence which did not seriously deviate from the actual experimental data.

In 1900, however, the German physicist Max Planck suggested a mathematical formula which would describe all the observed properties of blackbody radiation. Planck's formula was very good. It reproduced and predicted everything that the instruments measured with a very high degree of accuracy. It then became necessary to find out what the same formula said about the emission of light by atoms. Planck soon realized that what his formula said was, in effect, that the emission of light is not a smooth, continuous process. Atoms seemed to emit light in tiny, discrete, discontinuous jumps. This was a rather strange idea, and, at first, few physicists were prepared to believe it. They found it very difficult to imagine that natural processes could be discontinuous or jumplike.

It is easy to understand the reason for their disbelief. All our experiences with the world around us appear to show that changes are always continuous.* Consider this example: You push a freely hanging pendulum or a swing and it starts oscillating back and forth. These oscillations can be large or small depending on how forcefully you push. Imagine, however, that as you push this freely hanging swing, it starts to oscillate with a sweep of 1 meter or, if you push it harder, a sweep of 2 meters or perhaps 3 or 4 meters, but that it could never ever, no matter how carefully you adjust your push, swing with a sweep of 1½ or 2.657 meters—that when you increase the swing, it never goes up in other than integer increments. And correspondingly, that when you attempt to slow your swing down, it will slow down from a 3-meter sweep to a 2- or 1-meter sweep but will never slow down to a sweep of any other than integral length. A pendulum or swing would behave like this if, for some reason, it absorbed or gave up energy only in discrete, well-defined units and never fractions of units. Classical physics could never imagine such an object. Yet the assumption Planck used to explain his formula was that atoms emit light in the same manner as our imaginary pendulum swings

* This is likely to be a consequence of mammalian information processing. As we saw earlier, the mammalian senses tend to smooth out incoming information in order to make the world more predictable.

works. He assumed that atoms consisted of oscillating charges and that when they were hit by light, they changed the range of their oscillations and consequently changed their energy content. But these changes were not arbitrary. They had to happen discontinuously and there was a minimum amount of energy an atom could emit or absorb. This minimum energy was called an energy "quantum." In discovering the quantum character of the emission and absorption of light, Planck discovered that discontinuity is a fundamental property of nature and originated quantum theory.

In 1905, the same year that he created the special theory of relativity but preceding it by about three months, Einstein published a paper on a subject that had nothing to do with relativity: the paper was about the nature of light. Einstein suggested that light had a "corpuscular" structure: i.e., that it was composed of small concentrations of energy, or energy quanta (or, as they are called today, photons). The main reason that Einstein made this suggestion was that the wave theory, which at that time monopolized all thinking about light, could not explain how light ejects electrons from metal. (This phenomenon is called the photoelectric effect and is now widely used in photocells.) Einstein's hypothesis, on the other hand, explained all the observed properties of the photoelectric effect.

In his paper Einstein wrote that "the wave theory of light . . . has worked well in . . . purely optical phenomena and will probably never be replaced by another theory." Yet, he added in the next paragraph, some observations "are more readily understood if one assumes that the energy of light . . . consists of . . . energy quanta which are localized at points in space, which are without dividing, and which can only be produced and absorbed in complete units." In the rest of his paper he worked out some mathematical laws which followed from his hypothesis.

The subsequent fate of this paper is virtually unique in the history of physics. Consider the background: From 1905 to 1923, modern physics evolved rapidly and steadily. Theories which were at first controversial, such as the special theory of relativity, Planck's quantum theory, the first theory of atoms (known as the Rutherford-Bohr model), and Einstein's theory of Brownian motion and his magnum opus, the general theory of relativity, were all accepted into the mainstream of physics. These were, in other words, great years when revolutionary new discoveries became almost the norm.

Yet throughout this era of open, creative, and unconventional minds, of growth and experimentation, there was hardly a physicist who thought that Einstein's light-quantum theory was worth paying attention to. Such a slight is, of course, not unusual in the history of science when the author of a bold new theory is a beginner struggling for recognition. But Einstein was no beginner. During these same years, he was not only generally considered to be the leading physicist of the period, but was already acclaimed as the greatest since Newton. Yet his theory of light evoked nothing but embarrassment or derision.* In 1913, for example, Max Planck, together with three other eminent German physicists, approached their government with the proposal that Einstein, who was at the time professor of physics in Zurich, should be offered a special position in Berlin which would enable him to devote all his time to research. The proposal contained an evaluation of Einstein's contributions to physics and concluded: "All in all one can say that among the great problems, in which modern physics is so rich, there is hardly one to which Einstein has not brought some outstanding contribution. That he may sometimes have missed the target in his speculations as, for example, in his hypothesis of light-quanta cannot really be held against him." This was how the very founder of quantum theory, Einstein's great supporter and fatherly mentor Max Planck, among others, viewed the idea that light consisted of photons. And Planck and the others were not alone. Even those physicists who performed the very experiments which verified Einstein's equation of the photoelectric effect with high precision could not accept his premise. As Robert A. Millikan, an eminent American physicist, wrote in 1916: "Einstein's photoelectric equation . . . appears in every case to predict exactly the observed results. . . ." But he added that the "theory by which Einstein arrived at his equation seems at present wholly untenable." Einstein's theory of light remained essentially disregarded until, in 1923, the experiments of another American, Arthur H. Compton, unexpectedly yet finally and decisively proved its fundamental assump-

* Irrespective of the merits of this particular issue, this is a good example of how little authority often counts for in physics.

tions.* Compton's experiments showed that when light strikes a free electron, the process has all the characteristic properties of a collision between a particle like a "light-quantum" and a more conventional particle like an electron.

The only person who was not impressed by the universal rejection of the light-quantum hypothesis was Einstein himself. Not only did he not withdraw his theory, but he continued working on it and thinking about it intermittently and contributed important results throughout the years 1905–1924.

Why did Einstein's theory of light generate such strong, sustained, and uniform opposition? His theory of light was much more vehemently opposed than the theory of relativity ever was. It was not that Einstein wanted to abandon the highly successful wave theory. The quotation above makes this quite clear. But he insisted that light had "dual" properties—that it had corpuscular as well as wave properties. And here was the problem. It was his insistence on the dual character of light which made the theory impossible to accept. For it is simply not possible to imagine anything which is simultaneously (1) a wave extended everywhere in space and (2) a small, pointlike entity. It is like imagining that the earth is simultaneously round and flat. It makes no sense; it is not possible to think about light rationally with such a contradictory premise.

Einstein, of course, saw these problems as clearly as anyone else but he also saw, as in the case of the propagation of light in special relativity, that the facts of nature pay no attention to what we can or cannot imagine. Like it or not, light turned out to be both a "wave" and a "particle" simultaneously, and this peculiarity was not limited to light. Einstein's idea of the dual nature of light was only the first step in a conceptual revolution which turned out to be one of the most radical in the history of human understanding of the external world—only the first step, that is, in the development of the theory of quantum mechanics.

The two main actors in the early drama of the quantum were light and the atom. Einstein took care of light. The quantum nature of atoms took a longer time to understand in detail mainly

* When Einstein was awarded the Nobel Prize in 1921 for the photoelectric equation, the Nobel citation carefully ignored the photon hypothesis as it ignored his theories of relativity.

because their structure was still unknown. The first glimpse into this problem came in 1906 from the experiments of Ernest Rutherford, then of Manchester University. Rutherford discovered that atoms consist of (1) an electrically charged *nucleus* which is very small relative to the size of an atom, but which contains nevertheless virtually all the mass of the atom, and (2) lightweight electrons with a charge opposite to that of the nucleus which are at the periphery of the atom. It was assumed in this model that the electrons orbit around the nucleus. In 1912 and 1913, the young Danish physicist Niels Bohr, working at Rutherford's laboratory, offered the first credible theory of a mechanism which could explain both the stability of atoms and their emission and absorption of light. Bohr applied Planck's ideas of discrete energy states to Rutherford's model of the atom and came up with a theory in which electrons would stay forever on selected orbits around the nucleus and in which light was emitted or absorbed by the atom in a process in which electrons jump from one selected orbit to another. Bohr was able to derive mathematical conditions to determine these orbits.

The Bohr model was a giant step in the understanding of atomic mechanisms. It was the first atomic model which worked at all. It explained, for example, the highly complicated structure of light emitted by the simplest atom—the hydrogen atom—extremely well. No earlier atomic model had ever come close to solving this problem. Yet within a few years it also became clear that this theory could not be the final word on atomic structure. Although the Bohr model correctly described the behavior of the hydrogen atom, it was unable to explain the way the second simplest atom (helium) emitted light. Also, the model itself presented conceptual problems which defied understanding. Perhaps the most important of these was the problem of the selected orbits themselves. While one could write down formal mathematical rules for determining these orbits, there was no explanation in the theory as to why nature selected these particular orbits. It was also not clear what agents kept the electrons on these orbits. How did an electron "know" that these are the only orbits accessible? What prevented an electron from moving off its orbit?

A satisfactory solution of these problems was not far off. Quantum mechanics was about to be born. The history of this birth is quite extraordinary in several respects, and it is worth recounting, even if only briefly. Unlike many other great physical theories, quantum mechanics did not leap ready-made from the mind of a single genius. It was, instead, the work of a relatively large number of physicists. Among them were very young beginners as well as other older, more established figures. Some individuals worked from similar premises and followed parallel lines of thought. Others started from quite different, even contrary, ideas and proceeded in what seemed to be radically different directions. Sometimes they even disliked each other's ideas. A celebrated example: the Austrian physicist Erwin Schroedinger (one of the inventors of the theory) found the ideas of Werner Heisenberg (another pioneer) "repellent," while Heisenberg thought Schroedinger's theory "disgusting." There were other examples of lack of mutual comprehension. Yet the story had a happy ending. Not that the people involved came to appreciate each other's ideas. They did not actually have to. For it turned out after some initial confusion that, although they were working in seemingly different directions and using different techniques, they all ended up with *exactly the same theory*. This is what makes the early history of quantum mechanics so fascinating. It is difficult to think of a better example to demonstrate that the evolution of physics is much less determined by personal wishes, subjective likes and dislikes, cliques, personal and social relations than it is by the inner logic of facts and the ideas which explain the facts.

Quantum mechanics started in 1923 with the doctoral thesis of the French physicist Louis de Broglie. The thesis contained an idea which was as bold as it turned out to be lucky. Assuming that Einstein's theory of the dual character of light was right, de Broglie then went further and asked: Why limit this property to light? If light has dual character, if light can be both "wave" and "particle," then perhaps other entities which we had recognized as being particles only (such as electrons—or pieces of rock for that matter) could also have wave properties: properties which are hidden from simple observation but which might be (as, in fact, they subsequently were) discovered through skillful and careful experiments.* What encouraged

* These experiments were performed in 1927 in the United States (Davisson and Germer) and in Great Britain (Sir George Thompson) and fully verified de Broglie's assumption.

de Broglie was that this theory promised a solution to the problem of selected orbits.

About a year after de Broglie's work was published, the German physicist Werner Heisenberg worked out an ingenious mathematical method for describing the mechanism by which atomic electrons emit light. Heisenberg was interested neither in waves nor in particles. His ideas appeared to have no connection whatever with those of de Broglie. Heisenberg's main concern and chief achievement was to devise a consistent way of dealing with atomic mechanisms that avoided the use of concepts which did not correspond to directly observable quantities. He wanted to be able to work without them, concentrating instead on such properties of the atom which were directly observable, such as the frequency of the light it emitted.

Taken by themselves, neither de Broglie's nor Heisenberg's ideas, neither of which we can do justice to here, amounted to a "theory"; de Broglie had made an ingenious guess, and Heisenberg had produced a mathematical formalism. Both were shots in the dark inspired by earlier ideas: de Broglie's by Einstein's photon hypothesis, Heisenberg's by an important mathematical principle enunciated earlier by Niels Bohr. But both turned out to be very effective catalysts. In 1926 Schroedinger developed de Broglie's ideas into a more complete theory which became known as wave mechanics. Heisenberg's ideas were developed jointly by the Göttingen professor Max Born, Born's assistant Pascual Jordan, and Heisenberg himself and, independently and simultaneously, by a Cambridge graduate student, Paul Dirac, in 1925. In 1926, Schroedinger was able to prove mathematically that all these formulations were fully equivalent, and that the apparent difference between them was merely a difference in the mathematical language they used.

After 1926 quantum mechanics evolved in two directions. One was an "inward" direction; some researchers aimed at answering the innumerable questions associated with atoms, molecules, nuclear physics, chemistry, and solid-state problems, among others. In this direction, quantum mechanics turned out to be almost unbelievably successful. The theory could solve every single problem it addressed itself to, and nobody today doubts that in this domain it will remain successful in the future as well. The other direction was "outward," breaking into the then completely unknown quantum world of electromagnetism and of elementary particles. The trailblazer in this new direction was Dirac. In 1927, he extended quantum mechanics into the domain of electromagnetic fields, giving the first satisfactory picture involving the two actors in the quantum play, light and atoms. In another paper a year later, he established harmony between quantum mechanics on the one hand and the special theory of relativity on the other. One can say in hindsight that nearly all the work that has been done on the most fundamental problems of physics since that time had its roots in these two papers.

So much for history. In what follows I shall try to describe what quantum theory says about the fundamentals of human cosmology: about time, space, and the behavior of matter. An instructive way to approach this problem is to go into some detail about the "dual" character of the constituents of matter. All elementary particles display the "wave-particle duality," and, consequently, so does all matter. An electron (this particle will represent, in what follows, any elementary particle) sometimes behaves like a compact particle, localized in space, at other times like a spatially extended wave. All matter, in other words, displays the property Einstein first found in light. Electrons are both "waves" and "particles." But what is an electron in reality? A wave, a particle, some odd combination of the two, or does the electron change its nature from time to time?

To answer this question, it is helpful first to concentrate on one aspect of the "duality," consider that aspect in more detail, and later see how the other aspect could emerge. Let us forget about waves for a moment and concentrate on the particle aspect. How does the electron, as a particle, move? In classical mechanics, such a question could be answered exactly. If the *position* and the *velocity* of the particle and the forces acting on it are exactly known at a certain instant, then its future motion is also known exactly. Or, in simpler words, if you know where something is, how fast it is going, and how strongly it is pulled or pushed, then you can figure out where it will be after a certain time.

These data, however, cannot be obtained simultaneously when we are dealing with an elementary particle. Not even in principle. This important fact was discovered by Heisenberg, and this is the content of his famous "uncertainty principle." This is a subtle principle, and is as often and as thoroughly misunderstood as Einstein's principle of relativity. Before getting into the meaning of the uncertainty principle, however, let us consider a simple example of what a simultaneous measurement of the position and the velocity of an electron would yield if we attempted to perform it.

Imagine then that we want to determine the position of an electron by illuminating it in a microscope. We must first consider the following facts. Light has the fundamental property (this was part of Einstein's photon theory) that the shorter its wavelength, the larger the amount of energy it carries. It is also a fact that we can see an object only if it reflects light. An object, furthermore, cannot reflect light if the wavelength of the light is much larger than the object itself. Consequently, what will happen, roughly speaking, is this: If we use low-energy light, its wavelength is so long that the electron will not reflect it, and therefore the procedure will yield little information on the whereabouts of the electron. We can try to remedy the situation and use high-energy, that is, short-wavelength light. Then light will be reflected from the electron, and this can give precise information about where the electron was in that particular instant. But the impact of the high-energy light will kick the electron quite violently and uncontrollably from its position, so that we will get no information about the velocity of the electron at the precise instant for which we will gain information about its position.

The above experiment is important because it is characteristic of a situation and *not* because it reflects the human inability to determine these data simultaneously. The latter, by itself, would not be of such great interest. There are many things, after all, which we are unable to do. The point of this and other similar thought experiments probing the same question is that even if these experiments were to be performed by the finest tools *nature* (which includes all present and future human skills as well) can ever provide, there would still be no information available as to the simultaneous position and velocity of the electron. Not only can *we* not learn these data, but no computer, no supermicroscope, no ultrasmart demon of Laplace could ever do so. No matter what agents try to determine these data accurately, they will all be frustrated in exactly the same way, since subtler, more accurate measuring tools than electrons, photons, and the like do not exist. Consequently, exact information about the simultaneous values of the position and velocity of an electron does not exist in nature either. It does not exist for us or for any measuring device, nor does it exist for another elementary particle.

This sounds at first as if it must be an exaggeration. Surely the electron must be in some definite place, and, if it moves, it must move with some exact velocity.* Just because we cannot measure them simultaneously does not mean they can't exist simultaneously. Sounds reasonable, yet it isn't so. If, for example, we perform a calculation concerning the behavior of a particle in which we must assume a definite value for both the velocity and position, then we arrive at an empirically wrong result. Not only can we not measure position and velocity simultaneously, but we cannot even assume that they exist simultaneously. The following parable may help to explain the situation. Imagine a world where, by some quirk in the process of adaptation, the mammalian brain evolved with a system of complex nerve connections between the auditory and visual centers. This system of connections would produce the effect that every time we heard a sound of a certain pitch, we would see a certain color and vice-versa (the very sight of a color would create the sensation of a sound). From the moment of birth we would all perceive a livelier world where sounds had living colors and colors sounded loud and clear. In due course, we would also discover the empirical laws which determined what sounds evoke which colors and vice-versa. Only individuals with serious head injuries or birth defects would be without this ability to perceive the world in its full richness and would be condemned, for example, to hear voices without seeing colors.

But physics would also evolve, of course, and at one point some smart individual would discover that sound consisted of air pressure waves while light was nothing but electromagnetic oscillations. "Consequently," the smart person would say, "color and sound have nothing whatever to do with each other, and the connection that we sense between the two is merely an illusion that results from the particular way our nervous system processes auditory and visual information." Such an idea would seem at first

* It is momentum actually, not velocity, which figures in these arguments, but the distinction is unimportant for our purposes.

quite incredible since there would be nothing more obvious in that world than the fact that sound and color always came together, and that the laws determining how they came together were well-established laws of nature. But as more and more data accumulated, the new idea would come to be accepted and soon books would appear which would describe the new weird world of physics where colors have no sound and sound exists without colors.

The moral of the above piece of science fiction is obvious. The mammalian way of processing information from the directly perceivable world evolved in such a way that, independent of circumstances, we always endow every body simultaneously with an exact position and velocity. This may well be another inevitable by-product of the built-in editing process of our brain, which prefers to bring out the smooth, regular, and predictable features of the environment and tries to suppress the jerky, sudden, irregular ones. This tendency was probably very useful in adaptation, but that does not mean that this way of perceiving can make sense of reality at all levels. Quantum mechanics made us aware that, among other things, simultaneously precise values of position and velocity of particles do not exist. Thus, when we are given information from the microworld, we have to understand and make sense of it without all these data. The situation we are facing here is similar to the one we encountered in the special theory of relativity; there, the intuitively obvious idea of simultaneity and with it the absolute flow of time turned out to be mere conventions and not objective features of nature.

Yet even after we learn all this, the fact remains that no matter what elementary particles do or do not do, *our brain remains mammalian, and individual objects moving in a continuous manner in space and time is the only world it can ever really make sense of.* Therefore we are bound to "translate" for ourselves the nonmammalian and therefore to us very strange features of the microworld to a more familiar framework. It turns out that the way to do this in quantum mechanics is to replace a causal description of motion with a probabilistic one. Instead of describing in space and time the whereabouts of a particle, quantum mechanics offers a way to talk about the distribution, in space and time, of the *probability* of a particle being at some point.

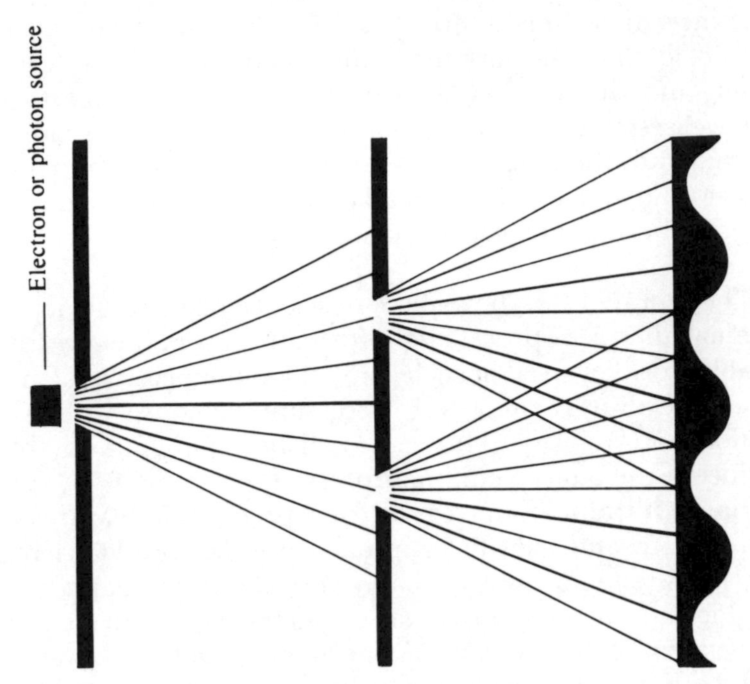

Figure 8-1. The two-slit experiment. Particles (electrons, photons or any other kind) coming from a source pass through two slits and hit a screen. Their rate of arrival is shown by the peaks and troughs behind the screen.

The following description of the so-called "interference experiment" (Figure 8-1) will make it clear how this is done. In this experiment, a large number of electrons are sent through two slits toward a sensitive surface, a screen, which records their arrival. (This surface can be a photographic plate where the electron leaves a tiny, dark dot on arrival.) We can send these electrons in one bunch over a short time or we can send them one by one taking a long time. The result is the same in both cases: The image on the screen is exactly the same as what we see when spatially extended continuous waves go through the same slits. The image consists of alternating dark and bright bands, called "interference fringes." Such an image is naturally expected when *waves* pass through two such slits. The waves recombine after passing through the slits, and where two wave crests meet, they reinforce each other, and where a crest combines with a trough, they cancel each other out. Hence the alternating bright and dark bands. This is a long-known, well-understood, simple process. It takes place only if both slits are open, since only in this case will the separated waves recombine. If we close one of the slits, all we find is a dark spot on the screen opposite the open slit.

In the nineteenth century, this type of two-slit experiment showed convincingly the wavelike nature of light, and it showed the wavelike nature of particles with equal force a century later. Other experiments can also be devised to find the unmistakable signs of the same behavior. Are electrons then in reality extended waves? They cannot be, since there is compelling evidence from other types of experiments that electrons are, in fact, indivisible particles so tiny that they have no measurable extension in space. The dilemma we are facing is, of course, not new. Einstein faced the same one when he postulated the existence of photons. But now this paradox shows itself to exist for electrons and, in fact, for every kind of matter. Quantum mechanics was developed in order to make sense of this paradox and it does so in the following way.

Let us go back to the particle picture for a moment. From this point of view, a dark band on the screen signifies one thing only. This is that more electrons arrived to form that band than to form a bright one. Electrons arrived all over the screen but more (many more actually) arrived in some bands than in others. A logical way to describe this is by saying that while the particlelike electrons could arrive at any place on the screen, they arrived with a much greater frequency, or the probability of their arrival was much greater, in a region where the plate was dark than in a region where it was bright. The interference fringes, the degrees of darkness, in other words, reflect the frequencies, the probabilities, the likelihoods, that electrons will hit the plate at a certain place.

This sounds reasonable and simple. What it implies, however, is that as electrons display wave properties by producing the dark-bright interference fringes, then these waves must be *waves of probabilities.* It implies that the motion of particles is associated with probabilities and that these probabilities propagate in space and time like waves. The significance of these waves is that through them we can interpret the motion of electrons as a motion in our perceivable world of space and time. We have to realize, however, that these are waves without a medium, waves made essentially from our concepts. Their properties are nevertheless measurable, and they are, in this sense, real. They display the space-time aspect of the motion of the electron.

The important discovery that the wavelike nature of particles reflects the way probabilities propagate in time and space was made by Max Born. Born thus created the all-important aid which made it possible for us to understand the behavior of elementary particles in terms of our inborn cosmology—i.e., as continuous processes in space and time. An example will help us to understand how the probability aspect works in practice. Consider Bohr's old theory of atoms. In this theory, electrons were assumed to orbit around the nucleus, forming a mini-solar system. The motion of the electrons in this now abandoned theory was clearly defined in time and space, and it was also assumed that electrons obeyed fully causal laws. Now we recall that Bohr's theory was a mixed success. In certain cases it worked very well, in others it did not work at all. This suggests that there were some features in this theory which corresponded to actual facts and there were others which did not. Using Born's interpretation, we can see the situation quite clearly.

Bohr's theory assumed that the one electron in the hydrogen atom moves on a well-defined orbit, and Bohr obtained good results from this assumption. Quantum mechanics, in contrast, postulates that the movement of the electron in the atom cannot be described by such a space-time concept as a definite orbit. It says, rather, that the electron can be anywhere with varying probabilities. But if now you use the rules of quantum mechanics and calculate these probabilities, then you find that these are very large around the points of Bohr's orbits and are quite small elsewhere. Thus the electron may, in principle, be anywhere, but, in fact, it will mostly be around the original Bohr orbit. This state of affairs explains why the idea of the orbits was successful and it also shows that the Bohr orbits do not exist at all.*

* In this particular example the superiority of quantum mechanics is not apparent. But it is far superior to the Bohr model for the following reason. The

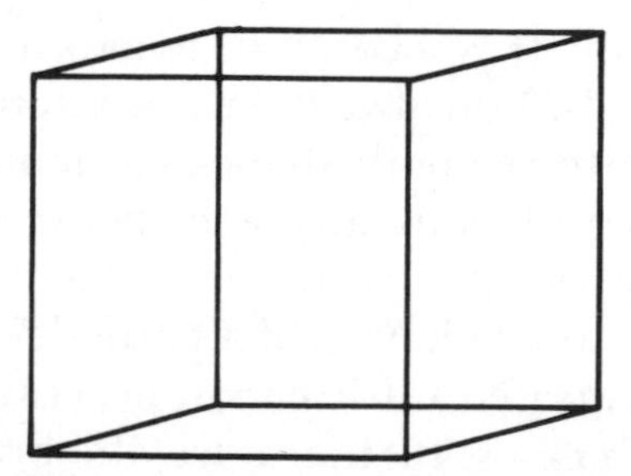

Figure 8-2. The Necker-cube. The image spontaneously reverses in depth.

In 1927, a year after Born's discovery of probability waves and in the same year that Heisenberg formulated the uncertainty principle, Bohr was able to define the type of information quantum physics can give to the human brain.* His celebrated "complementarity principle" states that we can either view the microworld in our mental framework of space and time *or* we can view it as being causally determined. But we cannot have both at the same time. Space-time description and causal description are complementary in nature and we can use only one of them at a time. Another way to express the complementarity principle is to say that we can observe matter either as particle or as wave, but we can never observe both properties simultaneously.

This principle expresses the essence of quantum mechanics. It also shows how far classical notions can be applied in the world of atoms. The message of this principle may be seen more intuitively by considering a parallel example which arises in visual information processing. The Necker cube (Figure 8-2) is a long-known visually ambiguous figure. If you look at it for some time, it re-

hydrogen atom has many observable properties. Bohr's model could explain a few of those. Quantum mechanics, in contrast, can explain every one of them. The Bohr model could not explain, for example, how two hydrogen atoms can form a hydrogen molecule. The picture of orbits was completely useless in this problem. Quantum mechanics explains molecule formation relatively simply, just as it explains every known fact in atomic physics.

* Bohr was the most ardent supporter of others' efforts to overthrow his own earlier theory.

verses spontaneously. If you continue to look at it, you will experience several reversals. You always see everything that is there, but you are never able to see both shapes simultaneously.

There are some obvious analogies between the two mental processes: "making visual sense of the Necker cube" and "making conceptual sense of the behavior of a particle." Both cases involve some highly sophisticated information processing within the human brain. In both cases, furthermore, the brain has two different models at its disposal. Each model corresponds to some aspects of reality, yet the two can never be employed simultaneously. We can use only one at a time. In each case, the question "Which model is the 'real' one?" is meaningless. Both models are "real"; the data support both. It is also meaningless to ask what the shape of the cube is when nobody looks at it. Without an observer, the problem originating from the nature of human information processing does not exist. The very act of observation, on the other hand, "forces" the object to take on one of the two possible appearances. This is as true for the electron as for the Necker cube, and this is why the role of observer is so important in quantum mechanics. Finally, the visual ambiguity does not impair our ability to collect all existing information about the cube. Nor does the principle of complementarity—or the uncertainty principle for that matter—prevent us from collecting all existing information about an electron.*

The complementarity principle is much more general than the uncertainty principle, but the two are related. The latter expresses quantitatively the extent to which the space-time concept of a smooth path is applicable to the motion of an electron. This is why the name "uncertainty principle" is a misnomer just as the tag "relativity" is. The principle says nothing about uncertainty. It

* The analogy between the cube and the electron is not between the two objects. One is, after all, a fundamental element of matter, the other a human construction aiming to expose the brain's inability to always process visual information in only one way. The analogy is rather in the way the brain processes information in both cases. In this context, it will not do to dismiss the cube by saying that it consists of twelve straight lines and nothing else. One could say with equal justice that the electron consists of a very well-known mass, charge, spin, and magnetic moment, and nothing else. Both statements are true, both are irrelevant to the real problem: How does our brain make sense of the information reflecting the appearance of these objects?

says, instead, that you can measure whatever there is in nature, but you cannot objectively measure what you only imagine to be there, what your inborn mental models project onto nature. It is, in fact, quite safe to say that the knowledge we can obtain about the microworld is in no way inferior to the knowledge we can acquire about the human-size environment. On the contrary. The mental discipline imposed by quantum mechanics, the rejection of preconceived notions about the motion of particles, endowed this branch of science with unbelievable accuracy and predictive power. There is no part of physics or any other science which could begin to approach the exactness of quantum theoretical calculations or the accuracy of quantum theoretical measurements.*

What turns out to be the case, however, is that in quantum mechanics ordinary space and time are not the best arena in which to interpret everything that happens. Even the spacetime of relativity is not always appropriate. More abstract structures, mathematical constructions are often more useful in making sense of the external world than space and time. This is the way the world is. The next sections use some examples to show this.

There is no known property of matter which is more significant and more profoundly important in determining the shape of the universe than the one expressed by the so-called "Pauli exclusion principle" or "Pauli principle" for short.† This principle was first formulated in 1925 in the framework of the Bohr theory. Pauli was studying the theoretical rules of atomic spectra (the patterns of light emitted by different atoms in their various states). His aim was to understand what determined the distribution of electrons among the "selected" orbits in the Bohr theory. He noticed that many observed properties of atomic spectra could be explained if one made the assumption that in any orbit selected by the Bohr theory there was never more than a single electron. This assumption was the first expression of the Pauli principle.

* Every physicist knows this. Yet popular books often neglect to point out this aspect of quantum mechanics. Perhaps uncertainties make for a more interesting story.

† Named after its discoverer, the Vienna-born Swiss physicist Wolfgang Pauli, a towering figure in twentieth-century physics.

In fact, Pauli was able to formulate his principle in a manner which survived the Bohr model. According to Bohr's theory, each selected electron orbit was characterized by four numbers. Each number had a specific significance. The nature of this significance,* however, is not as important as the fact that these four numbers completely characterized the orbit; i.e., there could be no two orbits with the same four numbers. These numbers were known technically as "quantum numbers." Instead of saying, then, that there is never more than one electron in a given orbit, the Pauli principle can be formulated to say this: There are no two electrons with the same four quantum numbers. This formulation is preferable because it does not refer to the imaginary electron orbits at all. Instead, it refers to quantum numbers only. These numbers remain meaningful in modern quantum mechanics as well. In quantum mechanics, they refer to a general and abstract, but nevertheless very exactly defined, concept: the *state*. It is a certain state which is characterized by quantum numbers in quantum mechanics. The Pauli principle then can be formulated this way: In a physical system *no two electrons can be in the same state.*

To appreciate the extraordinary significance of this principle in determining the properties of our world, all we have to do is to imagine a universe where it is not valid. Consider, to start with, what atoms would be like. Since electrons, like other physical systems, always tend to occupy the lowest possible energy state, without the effect of the Pauli principle all electrons in every atom would be in the lowest energy state. In the real universe, in contrast, only the simplest atom—hydrogen, which has only a single electron—has this property. Since the chemical properties of an atom are determined by the number of occupied states in it, without the Pauli principle at work all atoms would essentially be chemically equivalent to the hydrogen atom. There could be, in other words, as far as chemical behavior is concerned, no other element in the world but hydrogen. No carbon, no oxygen, no iron. Consequently no compounds, no earth, no life, no you, and no me. And this is just the beginning. The suspension of the Pauli principle would also play havoc with the properties of atomic nuclei and would make a tremendous difference even in such

* One number, for example, indicated how elongated a particular orbit was—i.e., how much it deviated from a circle.

large-scale properties of the universe as the life cycles of the stars. But the most awesome consequence of the Pauli principle dwarfs all others. As we shall see later, it is the Pauli principle which actually provides an answer, at least from the point of view of physics, to one of the deepest of all the riddles of philosophy: Why is there anything in the world instead of nothing at all? As we shall see, it is this principle which explains why all matter in the world does not disappear into vacuum.

The operation of the Pauli principle is not only all important, it is also very pervasive in nature. Not only do electrons obey the principle, but also neutrinos, quarks, protons, and neutrons (the last two themselves made up of quarks) among others.* And, finally, the physical effect this principle describes is extremely strong, probably infinitely strong. No physical action, however strong, could ever "squeeze" two electrons into the same state. The resistance against such pressure, for example, often sustains the weight of whole stars, some almost as big as our sun.

There is something remarkable in the way we speak about this principle. First, we call this tremendously important property of matter a "principle" and leave it at that. Second, this principle is seldom discussed either in popular or in philosophical terms. In fact, no matter what kind of book you open—technical, popular, or philosophical—the discussion of the Pauli principle usually never takes up more than a few pages. The reason for this sparse treatment is simple. There is very little one can say about this principle because the way it works cannot be described in spatial and temporal terms in the normal sense of these words. Consequently, ordinary language is inadequate to deal with it.

The reason that the action of this principle cannot be described as a spatial and temporal process is that we do not know of any physical agent whose action would produce the effect described by the Pauli principle. And this is the reason why we call it a "principle." If we called it an interaction, for example, we would have to imagine some agent carrying some kind of a force. We cannot do so. It is, in fact, virtually certain that no such agent exists separate from the electrons themselves and that effects due

* Particles which obey the Pauli principle are called "fermions" after the Italian physicist Enrico Fermi who first worked out the consequences of the Pauli principle for the properties of large numbers of particles.

to the Pauli principle have an entirely different origin. Whatever the origin, it does not correspond to anything intuitive. One thing we know about this principle is that all its effects can be described exactly by giving a mathematical object (called a "state vector") which describes the behavior of a collection of electrons (or other fermions) a specific structure. But this structure does not refer to anything spatial, nor does it depend on time in any way. It is nothing but an abstract property of a mathematical object.

Another thing we know about the Pauli principle, however, is that it says nothing indeterminate or probabilistic. It does not say that it is unlikely that there are two electrons with the same quantum numbers. It says, rather, that this cannot ever happen, period. The behavior of the electrons in this respect is exactly predictable, fully determined—and in this sense it is causal. Since what we have here is a strict law, we should not be surprised that it must be formulated abstractly without reference to space and time. For this is precisely what we would expect from Bohr's complementarity principle.

But if our good old mammalian brain still insists that it wants to imagine what happens between two electrons in observable space, we can also satisfy that desire. There will be, however, a price to pay, exactly as required by Bohr's principle. As we saw earlier, the use of space and time in quantum mechanics reduces certainty to probabilities. Suppose we now want to consider the effect of the Pauli principle on the spatial behavior of electrons. We can take the above mentioned "mathematical object" and use it according to the rules of quantum mechanics to find out, for example, how two free electrons (far removed from atoms and everything else) distribute themselves in *ordinary* space according to the prescriptions of the Pauli principle. What we find is this: Our two electrons can be virtually anywhere with equal likelihood (they are, after all, free) except that it is unlikely that they will be close to each other.* Here we see the complementarity principle validated again. When we wanted to know what happened in

* This property, incidentally, has nothing to do with the fact that, because they are both negatively charged, the two electrons repel each other. Chargeless fermions (such as the neutron) also obey this rule in the same way.

ordinary space, we were led to use probabilistic ideas.* We did not find that there was a strict law saying that two electrons *never* came close to each other but, rather, a probabilistic statement that it was unlikely they would do so.

As I said earlier, the Pauli principle explains a basic aspect of the stability of the world. To be more specific, we can use this principle to explain why we do not all disappear into vacuum. The notion that we actually should do so came up when, in 1928, Dirac successfully put the laws of quantum mechanics into a form which satisfied the requirements of Einstein's special theory of relativity.

Dirac was able to make the ideas of these two branches of physics compatible with each other by creating a new equation for electron waves—the celebrated Dirac equation. This equation is still the best available tool whenever we are interested in determining the behavior of a single electron under the influence of given external forces. It works very well in all such cases and, of course, this was the reason why this equation was readily and enthusiastically accepted by all physicists. But the greatest achievement of this equation came not through its many immediate successes but through what appeared to be its one and only failure. This failure gave rise to entirely new ideas which have had a great influence on what we now think about the nature of empty space, the direction of time, the origin of the universe—and even about such an old philosophical problem as the relation between the "actual" and the "potential." I will touch on some of these topics here.

The Dirac equation has the interesting property that, for every problem associated with the motion of an electron, it offers essentially two mathematical solutions. One solution describes the behavior of an electron in a way that is very much in accordance with what we observe. This is, of course, the solution which yields all those good results. The other solution, in contrast, describes

* It is important to realize that in its exact formulation, the Pauli principle does not say anything about electrons in ordinary three-dimensional space. It says that they cannot be in the same *state* and this concept has many characteristics. Spatial position may or may not be among them.

particles which have a very odd property: "negative energies." This is a nonsensical result for several reasons. First of all, no such thing has ever been observed in the universe. Second, if electrons could be in negative energy states, then matter could not exist in the universe in its observed forms at all. Take a hydrogen atom. If it was well isolated from external influences, the electron in it would be in its lowest energy state. If it was not in this state at the beginning, it would soon make a transition into it, since all physical systems tend to be in the lowest possible energy state. But if negative energy states existed, then there would be no lowest energy state, and the electron would forever make spontaneous transitions into ever-lower negative energy states! A hydrogen atom, therefore, could not exist in its normal observed state for any length of time. Nor could any other atom for that matter. In fact, we should not be able to observe any matter in the universe in its accustomed normal state. Everything would have to be in some unspecifiable negative energy state. Now this is so absurd that one is inclined to say that no matter how well the Dirac equation worked in solving certain problems, it surely had to be discarded or radically modified once it led to such absurd consequences.

Dirac, however, was convinced that his equation was correct and that since the negative energy solutions could not be explained away, they had to be meaningful. To find what they meant, however, was not going to be easy. He had to explain first of all why, if an infinite number of negative energy states existed, all electrons were not in these states. How, in other words, could positive-energy electrons exist permanently in the world? He soon came up with a bold, imaginative answer to this question. He theorized that what we think of as "empty space" or "vacuum" is, in fact, not empty at all, but completely populated with electrons occupying all available negative energy states. The so-called empty space of the universe, in other words, is really full of an infinitude of negative-energy electrons. These electrons, however, are completely unobservable. They do not exert any physical effect on anything in the world. The real world, including us, just exists in this "sea" of particles without ever feeling any of its effects, somewhat as we do not feel the tremendous pressure the atmosphere exerts on our bodies.

In addition to this infinite sea of electrons with negative energies there are, said Dirac, the real, observable electrons, which have positive energies. They cannot, however, enter into the negative energy states: *all these states are occupied and therefore the Pauli principle prevents other electrons from entering them.* Thus Dirac's idea explains at once why there are only positive-energy electrons in the world, and also why they do not all descend into negative energy states.

Of course, it is difficult to be impressed with such a theory when you first hear it. It borders on the absurd in many respects. It invents a whole new universe which cannot be seen or felt in any way. And we are asked to believe in its existence merely because it explains a wrong solution of an equation. But physics had evolved its own rules as to when a new theory could be accepted. Whenever one comes up with a new theory in physics, whether it sounds odd or quite credible, the crucial question is always the same: Can the theory *predict* things which had not been thought of before, and could those predictions then be tested by careful observations? Dirac therefore investigated the consequences of his hypothesis. Where did it lead? What new predictions could it make?

An important question which arises is this: Can we somehow perturb this infinite sea of negative energy electrons to make its existence manifest? Yes we can, because it should be possible for a negative energy electron to absorb a photon, i.e., energy from light. If the photon's energy is high enough, this energy can "lift" the electron into a positive energy state, thus converting it to a positive-energy, observable electron. A real electron, in other words. This can happen because the positive energy states are not all occupied and an electron can easily find one to enter. By supplying energy, therefore, we can lift an electron from its unobservable negative energy state and make it a decent, ordinary electron. But when such a process takes place, the vacuum remains short an electron; one negative energy state in it becomes empty. This fact can also be described by saying that by creating a positive-energy electron, by lifting the electron from its negative energy state, we are also creating an empty state or a "hole" in the vacuum. Sounds fascinating, but is it really what happens? Can you ever "see" such a "hole"? And if you can, how do you know it is a "hole"? The mathematical formalism of the Dirac equation is such that you can calculate the behavior and properties of such a hole in the vacuum and show that the hole will behave *exactly like an ordinary, positive-energy electron except that it will have the opposite electric charge.*

Thus the Dirac theory predicted that, with the help of energetic photons, you could create particles from the vacuum (i.e., quite literally from "nothing") which up to that time had never before been seen in nature—positively charged electrons or, as they came to be called, *positrons.* In addition, the theory says that when positrons come into existence, they must always be accompanied by ordinary electrons. (We cannot create a hole (positron) without lifting up an electron.) Finally, the theory also makes it clear that in normal circumstances, positrons cannot last long. They are, after all, nothing but "empty holes" and there are enough electrons in the real world to "fall into" these holes and "fill" them. When that happened, both the electron and the positron would disappear and an energetic photon would appear.

This theory, which certainly sounded farfetched at first, turned out to be one of the most successful in physics. Carl Anderson at the California Institute of Technology actually discovered the positron in 1932, two years after Dirac proposed his theory of the vacuum. Soon afterwards, all the other predictions of the theory were verified in laboratory experiments. Electromagnetic radiation indeed created electron-positron pairs from the vacuum, and these particles in turn annihilated each other whenever they met. And all these processes took place in exactly the way Dirac's theory predicted.

Thus not only was the only failure of Dirac's quantum mechanical equation rectified, but an entirely *new kind of matter* was discovered.* A kind of matter nobody ever dreamed of before which consisted of "holes in the vacuum" and annihilated ordinary matter every time it got into contact with it. This shadowy universe became known as "antimatter."† Nowadays positrons are used routinely in industry and medicine, and undergraduates

* We are not talking here about a new type of *material,* a new atom or compound, but, as emphasized above, about matter which is *different in kind* from ordinary matter.

† Some current theories predict other rather different types of shadowy universes observable at very high energies only. There is no experimental confirmation as yet, perhaps because existing accelerators do not have high enough energies.

repeat as a matter of course the original experiments with their at first quite unbelievable-sounding results. What boggled the mind in the 1920s became a commonplace only a few decades later. And, of course, it came as no surprise when from the late 1950s on, antiprotons, antineutrons, and a host of other representatives from the world of antimatter were gradually observed in various experiments.

In the late forties, the American physicist Richard Feynman pointed out that there was another interesting way to characterize antimatter. He showed mathematically that a *positron,* i.e., a particle having the same mass as an electron but carrying an equal and opposite electric charge, can be regarded as an *electron moving backward in time.* When he applied this idea to the Dirac theory of a single electron, it eliminated (for reasons too technical to get into) all the difficulties associated with negative energy states. There were no more negative energy states, there were only electrons that traveled sometimes forward, sometimes backward in time as they got scattered by some external agent like photons. Now traveling backward in time does not make much sense in ordinary macroscopic physics. On the contrary. The idea that time has one direction only is crucial in making sense of the directly perceivable world. Yet the idea turned out to be fruitful, led to no paradoxes, did not violate causality in any way and was equivalent mathematically to Dirac's theory of the vacuum. This fact indicates once again that the *ordinary* concepts of space and time are of limited use in the world of elementary particles.

We started out calling the completely matter-free regions of space "vacuum"—this is the common meaning of the word, after all—but ended up with the realization that vacuum was not empty at all. It contained as many electrons, protons, and everything else in negative energy states as you wanted to imagine. Yet in itself this whole collection of particles had no effect on anything—no gravitational effect, no electrical effect, no effect at all.

But the passivity of the vacuum lasts only until some kind of ordinary matter enters it—ordinary (positive energy) matter like electrons or photons or any other ordinary particles. It is in the presence of matter that vacuum starts to manifest itself.

To see what happens, we introduce another important so-called uncertainty principle. This is called the "energy-time uncertainty" and it was also introduced by Heisenberg. This principle says that if a particle exchanges energy with its surroundings, the range within which the amount of energy can vary is inversely proportional to the duration of the process. The longer the process takes, the smaller the range of the amount of exchanged energy.* For very short time intervals, according to this principle, the energy exchanged can fluctuate between wide limits, and, furthermore, the shorter the time span is, the larger the energy fluctuations can be. Consequently, the extent to which energy can be conserved in a process depends on how long the process lasts, and thus, for very short processes where the exchanged energy can vary between wide limits, and could therefore take on larger values, we cannot speak of even approximate conservation of energy. This is another phenomenon which is alien to and cannot be explained in terms of classical physics. Suppose now a photon propagates in vacuum. If the photon has enough energy to lift an electron out of a negative energy state, then (as described above) an electron-positron pair will be created. But according to the energy-time uncertainty principle, a photon can create an electron-positron pair even if it does not have enough energy if the pair exists for a very short time only. Even a not too energetic photon can turn into an electron-positron pair, in other words, if the pair annihilates a short time later and reverts again to a photon. It seems as if the photon were allowed to borrow energy from the vacuum for a very short time.

Particles which come into existence by violating the law of the conservation of energy and exist thus only for a very short time are called *virtual* in quantum theory. Not only photons can create a host of virtual particles. All other particles can do so. The virtual particles need not be electron-positron pairs either. There are many other possibilities.

* The actual numbers involved in this uncertainty principle, and in the one previously mentioned, are such that this restriction becomes completely irrelevant when macroscopic energy is exchanged during macroscopic time intervals. We never have to bother with uncertainty principles in ordinary life any more than we have to bother with the consequences of the absolute character of the speed of light.

There are two important facts about virtual particles. The first is that they can never be observed directly. And never really means never. They are unobservable in principle. Then why bother about them? Because of the second important fact: If we ignore virtual particles, quantum mechanics does not work. Here is an example. The neutron is a permanent constituent of atomic nuclei. It got its name from the fact that it is electrically neutral, i.e., has no electric charge. Yet it readily displays electromagnetic properties. This becomes understandable only if we realize that the neutron can, for very short time intervals, transform itself into electrically charged virtual particles. This is simply the way a neutron exists. The neutron, therefore, spends some of its time in charged states, giving rise to electromagnetic properties. And this is just one example out of many which show that if we ignore virtual particles because they are *unobservable,* quantum theory cannot explain the *observed* behavior of *observable* particles. In this sense, virtual particles can be observed indirectly through this influence.

We conclude then that a single particle—a photon, an electron, or any other particle—continuously creates a host of other particles by virtue of its very existence, and that all these particles create and destroy each other in dizzying succession. It is quite possible that what we call an "electron" is a manifestation of some very complicated fast processes taking place in the foamy structure of the vacuum. These manifestations are stable and display measurable properties like charge or mass, but they do not start to resemble a "billiard ball" or any other macroscopic visible model. That we feel tempted to imagine them as particles endowed with shapes, bodies, exact locations, and the like, is due to the inner workings of our brain. We would like to see even the microworld in terms of our own sensory models.

But this is not possible and it never will be. Quantum mechanics will not, because it cannot, offer intuitive mammalian models for the space-time behavior of particles. If we insist on having a sensory model for the microworld, we could do worse than borrow from the frog. For the frog's cosmology is a reasonably faithful caricature of the sensory models of quantum mechanics. In this cosmology, as we may recall, the world consists of shapeless, shadowy things which appear and disappear suddenly and un-

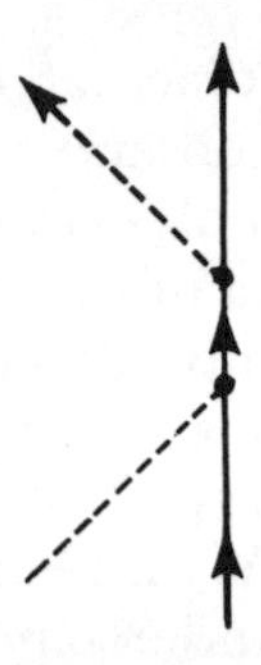

Figure 8-3

predictably. These things have no identity or individuality, no permanence in space and time. They can never be at rest. It is, in fact, part of the training of young physicists to depict such a world graphically in Feynman diagrams. These were invented by Feynman in 1948 as a visual aid in making calculations. An example of such a diagram showing the space-time skeleton of a process in which an electron scatters a photon is given in Figure 8-3. It is easy to imagine that animated versions of such diagrams would look familiar to an attentive frog.

Since the significance of such concepts as ordinary time, ordinary space, individual objects, and causality all diminish when it comes to elementary particles, ordinary language also becomes inadequate. Of all our important mental tools, it was only mathematics whose significance has not only not diminished, but has immensely increased in the strange world of the very small—and not only there but in all areas of physics. Galileo's remark that nature was written in the language of mathematics turned out to be more than an inconsequential bit of general wisdom. It has been consistently and thoroughly verified by experience during the succeeding centuries. So much so that mathematics seems to replace all other forms of human symbolism in the basic sciences. Thus the ever-increasing mathematization of science is a significant phenomenon which may one day throw more light on how the mind treats the external world. This subject may be worth a short detour.

To some extent, mathematics has always been associated with reality. Its development has always been fueled by problems which arose from observations of the external world. Yet since the axiomatization of geometry by Euclid, the logical consistency of mathematical rules and theorems has gradually become as important as, or more important than, affinity with or relevance to the actual world. During the last two or three hundred years, the independence of mathematics from the workings of nature and of sensory experience has become increasingly more pronounced. Mathematics became largely independent of the external world; it became the study of logical consequences and, in this sense, a study of the patterns of the human mind, of the human cognitive process. This is not to say that even today developments in physics or in biology or in some of the social sciences could not produce interesting mathematical riddles and keep mathematicians busy. But the bulk of mathematical problems have, for a long time now, been generated internally by the logic of mathematics itself. Nothing in the real world, to give a simple example, has ever demanded the study of complex numbers. No practical problem ever required that we known the "value" of $\sqrt{-1}$. But in the eighteenth and nineteenth centuries, mathematicians were already developing not only the theory of complex numbers, but also a magnificent framework of complex analysis. This elegant and powerful method, in turn, became extremely useful in the theory and applications of electromagnetism.

This well-worn example illuminates the curious relationship between mathematics and physics, which goes much beyond the fact that the former is the "language" of the latter. Physics, after all, collects information from the external world by performing measurements, and measurements mostly result in numbers, so it is not all that surprising that mathematics is the best method for expressing whatever pattern the measurements show. Nor would it be surprising that mathematics became capable not only of expressing but also of interpreting, extending, and generalizing the patterns we found in measurements *if* the mathematics used had been developed from problems from the real world and thus reflected the properties of that same world.

The intriguing aspect of the partnership of physics and mathematics lies in the fact that even such parts of mathematics as were developed without the slightest reference to anything in the real world often turned out to be ready-made for a perfect description of some physical situation. Eugene Wigner, the distinguished theoretical physicist of Princeton University, called this phenomenon "the unreasonable effectiveness of mathematics in the natural sciences."

Much of mathematics never gets used in physics or elsewhere. But until now—and there have been very few exceptions—when physicists needed new mathematical tools to develop new physical theories, they found these tools ready and waiting in some earlier mathematical work. These tools, in other words, were developed in contexts which had nothing whatever to do with the problem to which they were later applied. Quantum mechanics, for example, could not have developed without the use, among other things, of the theory of certain infinite dimensional spaces. Just such a theory was developed in mathematics in the two decades preceding the advent of quantum mechanics. The general theory of relativity could not have been formulated without a detailed mathematical method capable of treating the geometry of curved Riemannian spaces. Such a method was developed from 1880 on by Italian mathematicians who, in turn, had never thought that what they were doing could ever have any relevance to the problem of gravitation. Or take the example of the elementary particles, most of which were discovered during the last thirty-odd years. These particles showed some abstract patterns in the way their masses, charges, and some other properties were distributed. These patterns made sense only when group theory—an abstract branch of mathematics which evolved during the last two to three centuries to a high degree of completeness without having any contact with physics—was invoked.

One could find a great many such examples. All would show that so-called pure mathematics, while motivated exclusively, as Wigner puts it, by "ingenuity and the feeling of formal beauty," displays an uncanny ability to describe even properties of the real world which were completely hidden and unknown at the time when the mathematical tools were actually worked out. And there is something even more amazing. In sharp contrast to the natural sciences themselves, the more mathematics disregarded sensory impressions and intuitive mental models, the more suitable, the more effective and powerful it became as a tool for describing nature. We find all this difficult to explain. We will have to know much more about the subtle interplays between the external patterns the real world imposes on our nervous system and the internal patterns our nervous system imposes on the real world. Lacking such knowledge, we have to be satisfied with Galileo's metaphor.

9

Creating New Realities

(Space and Time in the Arts)

If nature were not beautiful, it would not be worth knowing and life would not be worth living . . .

Henri Poincare

The sudden realization of the inadequacies of the cosmology we are born with and the crisis this realization caused at the turn of the century were not confined to physics. In the last two chapters, when we considered the evolution of physics only, it seemed an obvious conclusion that the failure of our intuitive expectations to fit the external world was due exclusively to our explorations into the unperceivable extreme dimensions of the large, the small, and the fast. But the actual historic and mental processes which led to the demise of the mammalian structure of our worldview seem to be much more complex.

At the very same time that modern physics was first encountering and dealing with hitherto unimaginable concepts, the expressive forms of the arts, painting and music in particular, were also breaking out of the classical mold—the framework of time and space, of motion and order. By the early part of the twentieth century, the "normal" perceptual models of reality, which for centuries had served the arts so well, no longer seemed to be sufficient.

Although radically new ideas about the nature of the world emerged simultaneously in the sciences and in the arts, these processes were by all accounts completely independent of each other. They were motivated by different concerns and carried out through different methods by people who knew nothing about each others' problems or results. Yet there may be some connections between these events even if we cannot pinpoint them. Both science and art are, after all, expressions of the human imagination; the external world, furthermore, as well as our brain, appears to contain certain patterns, and the artist and the scientist both look for those patterns which are significant; they look for patterns which help make sense of the world. Seen from this point of view, the simultaneous failure of the mammalian cosmology to be useful in physics, in music, and in the visual arts seems less strange but no less interesting. In what follows, after some brief detours into past history, I shall discuss the changes in the notions of space embodied in early twentieth-century art and the changes in the representation of time and motion in music.

We saw earlier, when we described the difference in the spatial concepts of Egyptian and Greek sculpture, that the conceptions of space of a particular society are reflected in and probably influenced by the arts. We noted this also when we discussed the Italian Renaissance, seeing how the discovery of the mathematical rules underlying our sense of vision introduced the era of classical space. We see now in retrospect that the Renaissance was only the beginning of a long, fertile, and wide-ranging evolution in the visual arts which lasted almost five centuries. Once the technique of perspective for depicting depth had been mastered, the focus of painting shifted towards an ever-increasing exactness in depicting the world our senses inform us about. This included the varieties of texture and shades of color, the inexhaustible effects of light and shade, human faces and figures reflecting an infinite variety of feelings, of emotional and mental states. In the course of this evolution, virtually every decade, every country, and every influential artist developed new ways of representing the human experience of vision, of space, of form, and of the emotions associated with them. Each age, each artist, used characteristic means, and the results often bore little resemblance to the works of other

Figure 9-1

artists. But as different as they may have been, they all had one feature in common: They were all representations of the dominant spatial concept of Western society, classical space. From the early fifteenth to the late nineteenth century, every painting, every sculpture, depicted the world, whether real or imaginary, as consisting of separate, well-defined, and recognizable figures in extended three-dimensional space.

This basic perceptual framework remained the same throughout five centuries, and only this framework provided the continu-

Figure 9-2

ity between the works of such otherwise totally dissimilar artists as, for example, Diego Velázques in the seventeenth century and the Impressionist painter Claude Monet in the late nineteenth. (Compare Figures 9-1 and 9-2.) Nor did this perceptual framework change fundamentally with the artistic concerns of the painter. Rembrandt van Rijn, for example, in the seventeenth century, to quote from the widely known Gardner's *Art Through the Ages,* "made painting a method for probing the states of the human soul." Paul Cézanne, on the other hand, at the end of the nineteenth century, was preoccupied with looking for what he thought to be the essential spatial structures underlying an ever-changing visual impression. Yet neither Rembrandt nor Cézanne ever painted a picture which depicted anything but three-dimensional space containing individual, recognizable objects in well-defined places. Throughout these centuries, one way or another, the symbols of artists, no less than those of the physicists, were essentially consistent with the mammalian view of the world.

It was precisely this basic sensory framework, the depiction of the world in a recognizable way, which changed at the beginning of the twentieth century as abruptly and as completely in painting as it did in physics. In the years 1906–1908, the first of the modern artistic movements, the so-called cubist school of painting, came into existence. The method was created by Georges Braque and Pablo Picasso, and further developed by Juan Gris and others. While the strictly cubist way of painting did not last very long (it remained influential until about 1920 or so), the influence of the cubist movement on the theory and practice of painting remained great and irreversible.

Cubism was devoted to the creation of a visual world which was not only new in painting but was also very different from what ordinary vision takes in. These artists were interested in analyzing the permanent components of the visible world. Since human vision cannot really perceive depth, only depth cues, the cubists rejected the three-dimensional illusions of earlier paintings and concentrated on two-dimensional surfaces. They created a new visual order in their pictures. Unlike some earlier painters, whose chief concerns were with the depiction of human emotion, the cubists' focus was entirely on the elements of vision. They decomposed the appearances of objects into visual surfaces and reorganized them in various manners; they broke down visual impressions and then reassembled them in new ways. As a result their pictures often looked, as the art theorist John Golding writes, as if the painter "had walked 180 degrees around his subject and had synthesized his impressions into a single image. . . ." Figures 9-3 and 9-4 show celebrated cubist painting by Georges Braque and Pablo Picasso.

Cubism was the first total break with the Renaissance traditions in Western painting, and for this reason alone the cubists were some of the most radical revolutionaries in the history of art. Cubism was the first artistic school which established the independence of painting from what was immediately visible. And, paradoxically, it was precisely by ignoring visual appearances that cubism opened up new possibilities of representing visual reality.

Figure 9-3

Figure 9-4

Current theories of vision assert on the basis of experimental evidence from brain research that an important part of visual information processing consists specifically of the breaking up of direct sensory input into components such as edges, angles, and straight lines of varying directions, and reassembling these into an inner picture of what we eventually "see." Some of the basic components of sensory input, in other words, are not the ones that we are conscious of in seeing. The final picture is the result of a highly complex mental process. In their search for visual laws underlying appearances, the cubists seem to have appreciated this intuitively before anybody else.

But cubism was just the detonator of an explosion: there followed an extremely rich period of searching and experimentation in the visual arts, all concerned with questioning our previously accepted perceptual modes. Twentieth-century art branched out into several important directions; a number of new schools began aggressively challenging virtually all the visual aspects of the human cosmology. Some of these schools were particularly interesting. The so-called "surrealist" movement, for example, came into existence in the early 1920s and had important adherents as late as the 1960s. Different manifestations of this movement were to be found in several art forms, including literature, and all were strongly influenced by the psychological discoveries of Sigmund Freud. Surrealism used a variety of techniques but was held together by the basic belief that certain parts of reality can only be known if we rely on our instinctive, subconscious experiences. Painters of this school, among whom the German-born French artist Max Ernst and the Spanish artist Salvador Dali were particularly influential, tried to make our dreamlike subconscious fantasies visible.

In order to challenge the trust we place in our sensory experiences alone and turn our attention towards a kind of "inner" vision, a surrealist painter may, for example, use the following technique: The painter will depict individual objects or sights carefully, realistically, and with great accuracy, but will juxtapose these objects in illogical, often deliberately irritating ways, confusing and questioning the meaning of what we see. This dislocation and lack of visual logic startles the viewer and produces a feeling of unreality which contrasts sharply with the realistic elements and background of the picture. The result is that we often feel that scenes in surrealistic paintings are similar to those we see or think we see in our dreams. Such effects are already marked in the paintings of a forerunner of the surrealists, the Italian painter Giorgio de Chirico.

Many of the surrealist works were also meant to stand as statements of a kind of visual philosophy. We saw in the previous chapter, in the example of the Necker cube, how visual analogies can help illuminate perceptual problems. Two artists of the surrealist school whose work pointed to the difficulties inherent in the separation of the world strictly into the "external" and "internal" and who used visual ambiguities to question the role of vision in human perception, were the Belgian painter René Magritte and the popular graphic artist who often used mathematical ideas to create surrealistic pictures—Maurits Escher of Holland. An expressive example of this type of later surrealistic painting can be seen in the 1961 painting by Magritte shown in Figure 9-5.

But the artistic movement which was the leader in rejecting our reliance on direct sensory experience—and which in this re-

Figure 9-5

gard was analogous to pure mathematics in the realm of the sciences—was the one misnamed "abstract" art. The term is a misnomer since all art is abstract, symbolic. But, once again, the name stuck and we live with it. This art is perhaps the most characteristic of the twentieth century. In abstract art, the artist completely disregards the "preexisting" visual world and creates totally new visual forms, forms, in other words, which did not exist earlier and which carry no reference to, no information about anything in the world outside of the work. This is the kind of work which the French sculptor Raymond Duchamp-Villon refers to by saying: "The sole purpose of the arts is neither description nor imitation, but the creation of unknown beings from elements which are always present but not apparent."

In this type of art, painting or sculpture acquired the same radical independence from the external world that mathematical thinking had achieved much earlier. Both were less the reflections of external reality than of the creative groping of the mind. In 1973, the art critic Patrick Heron described abstract painting and its role thus:

> Painting's role in civilization is that of man's laboratory for the disinterested exploration of visual appearances as such, an exploration carried out uninhibited by any practical demands whatsoever. The painter is and always has been in search of one thing only: and that is a new abstract configuration, a new but purely formal significance, a new pattern emerging out of the very mechanics of physical vision itself. . . .

If you just change a few nouns in this sentence, replace "paintings" with "mathematics," "visual" with "logical," and so forth, it could serve quite well as a definition of pure mathematics. And how does one evaluate abstract paintings? "Beauty and seriousness were the criteria by which his patterns should be judged." This sounds perhaps a bit academic, but it would do as an instruction to would-be art critics. It was written by the English mathematician G. H. Hardy, who was merely explaining how one judged a work in pure mathematics.

Abstract art took on a number of different formal characteristics. One school, sometimes called "constructivism" or "geometri-

Figure 9-6

cal abstraction," made use almost exclusively of well-defined, very regular geometrical forms, such as straight lines, angles, colored shapes with straight boundaries, to formulate their ideas. Works of this type may well have been stimulated by earlier cubist ideas. These artists did not aim to express deeply felt emotions or to depict the worlds of the subconscious. Their aim was rather to stimulate and educate human perception. The painters and sculptors of this school also thought that the simplicity and clarity of their constructions would make them more objective, more accessible, and thus more universal. Figure 9-6 shows a famous painting by the Dutch painter Piet Mondrian, one of the pioneers in the use of geometric abstraction. The leading artists of this movement, among whom were the painter Kasimir Malevich of Russia, the Russian-American sculptor Naum Gabo, and the versatile Hungarian-American artist László Moholy-Nagy, exerted a powerful influence not only on other artists but on the entire visual character of our contemporary world. The influence of their ideas extended everywhere—from architecture and town planning to the design of household articles.

Another influential school in abstract art became known as "expressive abstraction." The artists of this movement had perhaps more in common with the surrealists than with the cubists. Instead of simple, universal geometric shapes, they created irregular, colorful, complex forms. These forms evoked the often disorderly appearances of living things while paying scant attention to images of simple order. The shapes, the colors, the lines, the planes, and the materials were all chosen not for their orderly structure but for their expressive potential. These works often possessed organic or sensual, sometimes erotic, qualities; they aimed at expressing the unique, the subjective, the individual; the unconscious more than the conscious in the human mind. The Russian-born Wassily Kandinsky and the Spanish painter Joan Miro were among the most celebrated artists of this school. A later (1940s), influential offshoot of this school was the American movement known as "abstract expressionism" which originated in New York. Abstract expressionism took the impulsive spontaneity of expressive abstraction to its logical conclusion. The best known of the abstract expressionists was the American Jackson Pollock, who created his most famous paintings in an improvisatory, intuitive manner by dripping paint on a canvas placed flat on the

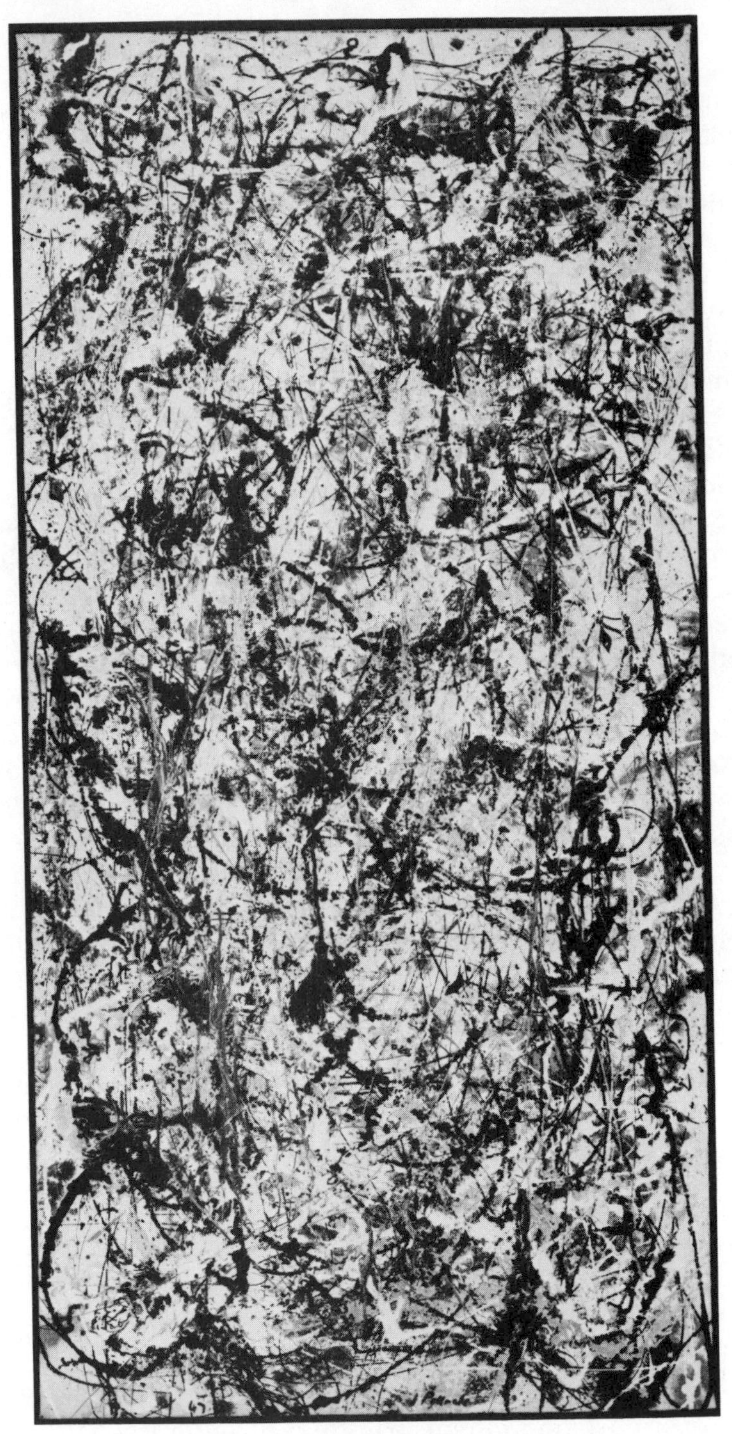

Figure 9-7

ground. These paintings look chaotic, yet they are able to communicate an inner pulsation and excitement. They are far from being realistic, but they are not accidental either. An example of an image of such meaningful and inspired disorder is given in Figure 9-7.

A question arises quite naturally at this point. We saw earlier that the spatial concepts of classical physics and the visual space of painting as it evolved from the Renaissance onward were compatible symbolic spaces. They were characterized by the same properties, and both seemed compatible with what most human beings then felt to be the "normal" perceptual space. We could call both by the common name of classical space. But what about the "space" of twentieth century physics and that of twentieth-century art? Can we say that the space of relativity or quantum theory corresponds in any way to the space of a cubist painting or a constructivist sculpture?

The answer seems to be a qualified "no." The twentieth-century notions of space, as we saw earlier, did not evolve from direct, immediate sense perceptions, either in science or in the visual arts. These notions were the results of searches into what may be hidden beyond superficial appearances or immediate impressions. This was perhaps the only common element in all the revolutionary changes at the turn of the century. The problems faced were very different in physics and in the arts, and the newly invented symbolic spaces, and the answers to these problems, were also different. There were various suggestions in earlier literature, dating from the 1930s and 1940s, that special relativity and cubism actually employed similar notions of space. This can (as Einstein himself once pointed out) hardly be the case. The spatial (or, rather, the space-time) notions of relativity evolved logically from the demands posed by the constancy of the speed of light. The problems of cubist artists had nothing to do with this problem or with mathematically formulated logic. The cubist space, for example, tended to be a two-dimensional surface which excluded the third; the mathematics of relativity works in four-dimensional space-time. And one can continue indefinitely; looking for similarities in these two enterprises is quite useless.

Does this all mean that the contemporary system of symbolic space and time, which should frame the general worldview of our society, is now hopelessly split into mutually alien artistic and scientific systems? Superficially this seems indeed to be the case. However, according to the art theorist Leo Steinberg, this viewpoint, which assumes that a strict separation of the functions of the "mind" and the "eye" is possible, is quite wrong. "The eye is part of the mind," wrote Steinberg in an often cited study in which he argued that modern painting in general and abstract art in particular has not given up exploring nature but turned instead toward studying and representing its often hidden fundamental features. Steinberg argued furthermore that while it was true that science no longer studied the directly visible, vision nevertheless will continue to play a fundamental role in the totality of the human cosmology. Arguing with those who claim that the role of art in exploring the world has become unimportant in the modern world, Steinberg pointed out that their arguments . . .

> contain an unwarranted assumption, to wit, that whereas man's capacity for intellectual abstraction is everwidening, his visual imagination is fixed and circumscribed. Here the philosophers are reckoning without the host, since our visualizing powers are determined for us not by them but by the men who paint. And this our visual imagination, thanks to those in whom it is creative, is also in perpetual growth, as unpredictable as the extension of thought.

In 1958, George Schmidt, a Swiss art historian and director of the noted Kunstmuseum in Basel, organized an exhibition entitled Form in Art and Nature, which can be considered as an interesting and specific illustration of Steinberg's point. The exhibition presented a number of paintings, some purely abstract and all from the twentieth century. Hung alongside the paintings were photographs of microscopic images—some taken with the electron microscope—of a number of organic and inorganic forms. Great similarities between the paintings and the photographs were immediately apparent, although the photographs were all taken long after the paintings were done. This may sound like a trivial coincidence at first. And obviously I do not mean here that the artists in question had some prophetic premonition of the particular shapes appearing in the later photographs. Presumably, for every photograph chosen for the exhibition, there were thousands of others with shapes not represented in the paintings. But the visual appearance of microscopic objects like stained cells in human organs or crystalline structures of inorganic materials are very different from those found in our everyday surroundings. Therefore, no selection of nineteenth-century (let alone earlier) paintings, which always re-created in one way or another the directly visible, could possibly have shown similarities with submicroscopic pictures. The abstract painters, in other words, have enhanced the visual vocabulary of art to an extent that only extraordinary and previously inaccessible shapes could be matched against the paintings.

But more than any specific examples, our whole world testifies to the fact that the visualizing powers of our brain have certainly not ceased to grow, that the visual exploration into the unknown has not ended. On the contrary. The new ways of painting have opened up a previously unseen and unknown new world of vision. But this world is not the classical world. It is genuinely new, it has never been seen before. Not only has our mental world in general been enriched, but there is reason to speculate that every aspect of our thinking, including the cognitive, has been profoundly influenced. Spatial thinking in general and visual thinking in particular, for example, seem to be of the greatest importance in the thought processes of active and creative scientists and mathematicians. This point was made convincingly by the distinguished mathematician Jacques Hadamard with a detailed study of the psychology of mathematical-type creativity, and has been supported by ongoing psychological research. In addition, we also have the personal testimonies of such great physicists as Einstein, Dirac, and Feynman, all stressing the importance of visual images in their thinking.

There is also evidence that the development of visual ability and imagination is greatly influenced by the visual environment in which we live, by the visual sophistication of the society of which we are members. Now twentieth-century art evidences a degree of visual sophistication which never existed before. Since it influences cognitive thinking, even if through the subconscious, one can even speculate whether the frequency and the scope of the epoch-making discoveries in the basic sciences would have been the same if the visual imagination of our society for some reason could not have kept pace, in sophistication, in the ability to abstract and to generalize, with our cognitive processes. The answer to that is probably "no." Our mind is an integrated whole and it does not seem possible to imagine a naturally evolving society which thinks in terms of relativity and quantum mechanics, yet at the same times sees the world in the visual frame of the nineteenth century.

As we have seen before, as much as the visual arts shape and reflect the space in which humans imagine the world, music tends to form and mirror their notions of time. This was the case at least in the last several hundred years of Western cultural evolution. As we know, polyphony and its notations literally created metric time in Europe. Music from the seventeenth century to the end of the nineteenth reflected, as we shall see later, the classical notions of time, motion, and causality. And again, in the early years of the twentieth century, as our inborn cosmology was being challenged in physics and in the visual arts, similar radical changes in our feelings about time were reflected in music.

Music seems to be necessary for human life. We do not know of any society without some kind of music. Thinkers as different as the American biologist Lewis Thomas and the French anthropologist Claude Lévi-Strauss have suggested that key elements in the workings of the human mind and thus in the human cosmology can only be understood through an understanding of how the mind creates and reacts to music. Music also has been many things to many people. In most societies, it was assumed that it possessed magical power. But it was also in music that mathematical order was discovered repeatedly in many civilizations. Nowadays, the universe is sometimes thought of as a kind of supercomputer, where the laws of physics are the software. In a similar vein, since the time of the Pythagoreans and through the Greek-Roman-Western evolutionary line, the universe was imagined as a giant musical instrument playing its own melodies, melodies in which the secret mathematical laws of the world were to be found. The fifth-century Roman statesman and philosopher Boethius suggested that there were three kinds of music, each with its own role in keeping order and harmony in the universe. The first, *musica mundana,* which was akin to Pythagoras' "music of the spheres," was responsible for the large-scale order and coherence of the universe. This music, although inaudible to human ears, was embodied in the motion of the heavens; it organized time and regulated the seasons and like phenomena. The second, the equally inaudible *musica humana,* was necessary for keeping the human body and soul working together in harmony. Finally, *musica instrumentalis,* i.e., audible music, while not as important as either of the others, was nevertheless also able to express order and structure, and, in addition, it gave pleasure to the listener.

Throughout medieval times and the Renaissance, the theory and practice of music was considered to be a science, the study of which was necessary for an understanding of the structure and the workings of the universe. The laws of the music of the time were considered the basic laws of nature, and this idea survived up to the beginning of early modern science. Johannes Kepler, for one, was convinced that the motion of planets obeyed musical rules expressible by mathematics, and it was this assumption that motivated him to look for laws of planetary motion in the first place. The scientists' feelings were echoed by musicians and music theorists. If the mathematical rules of nature were derivable from music, then the rules of music must be considered applied mathematics. When the most influential of all Renaissance music theorists, Gioseffe Zarlino, discussed polyphony, he did not concern himself with a composer's "self-expression" but delineated the character of his subject right from the start: "Every mathematical science relies upon demonstration rather than argument and opinion." He was entertaining no doubts that music was a mathematical science. About two hundred years later, when the French composer Jean-Philippe Rameau codified the rules of harmony and laid down the rules of music theory that were to remain in force for another two centuries, the title he chose for his work, *A Treatise on Harmony Reduced to Its Natural Principles,* was clearly meant to evoke that other, then forty-year-old, influential book by Newton, *Mathematical Principles of Natural Philosophy.* Another two hundred years later (in 1971), but in much the same spirit, the pianist-musicologist Charles Rosen summarized music's perceived role in Western history:

> Since the Renaissance at least, the arts have been conceived as ways of exploring the universe, as complementary to the sciences. To a certain extent, they create their own fields of research: their universe is the language they have shaped,

whose nature and limits they explore, and in exploring, transform. Beethoven is perhaps the first composer for whom this exploratory function of music took precedence over every other: pleasure, instruction and even, at times, expression. A work like the "Diabelli" Variations is above all a discovery of the nature of the simplest musical elements, an investigation of the language of classical tonality with all its implications for rhythms and texture as well as melody and harmony.

Even more radical was the opinion of Igor Stravinsky. He denied that music could ever actually express anything and said that its *exclusive function was to structure the flow of time and keep order in it.** Be that as it may, Stravinsky's statement reflects what is perhaps the most profound and crucial of music's many roles in the human cosmology: that it mirrors and affects our feelings toward the passage of time.

Earlier, when we mentioned polyphony and its effects on the notions of time of an entire civilization, we were quoting a specific example—one in which the connection between music and time was at its most obvious. But music's influence on the perception of time is probably much more general because music represents a most important abstract image of motion. This seems to be a general property of all music. Human beings always feel that music moves: It may move quickly or slowly, accelerate or slow down or even just "keep the pace." People even speak of music as progressing and think of it as rising or falling and so on. This motion is, however, purely symbolic. It is no more real motion than the space in a realistic painting is real three-dimensional space. It is "motion" for the human mind only. In reality, there is nothing in music which actually moves (the vibration of air accompanying any sound has no relation to the intrinsic musical motion of tones). Real motion always involves time and space. The symbolic motion of music has no spatial components and no spatial dimensions at all. It is a purely temporal phenomenon. It is "motion" which occurs only in the dimension of time.

The motion in music was always considered mysterious and

* The feelings and emotions that music arouses are, in other words, those of the listener and not of the music or its composer.

many had asked in the past "what" it was that actually moved. Victor Zuckerkandel pointed out that in each historical period, musical motion reflected what that period considered important. In the seventeenth and eighteenth centuries, when the aim of all art was considered to be the true imitation of nature, music was viewed as imitating the motion of feelings (whatever was meant by that phrase). The romantic era associated musical motion with some mysterious flow of life. At the end of the last century, the Austrian music critic Eduard Hanslick defined music somewhat less mysteriously as being "sounding forms in motion." In considering motion, we should remember that it is also part of our nervous makeup that we can only perceive motion relative to something fixed, be that a fixed point or object, a whole background, or anything else. We cannot perceive nor can we imagine motion without attaching to it a state of rest. This applies to symbolic musical motion as much as it applies to real physical motion. For musical motion to unfold, it also needs a representation of rest. Such a representation of rest can be found in all known music in all known societies.

All music expresses rest by means of what is called "loyalty to a central tone." The principle of loyalty to a central tone seems to have been obeyed by every kind of music everywhere in the world. It was observed in non-Western civilizations, in Gregorian chants, in polyphonic music, in all classical and nonclassical musical forms. The Beatles obeyed it and so did J. S. Bach. There is nothing simpler than to demonstrate the principle of loyalty to the central tone and its connection with the feeling of rest in musical motion if one considers the basic scale (since the eighteenth century) of European music. Simply start singing the "do-re-me-fa-so-la-ti-do" that we all learned in school. No music theory is needed to feel upon reaching "ti" a necessity to go up to "do" and finish there. Stopping at "ti" or at "la" creates a feeling of unfinished motion, a tension which wants to be resolved. The same effect can be felt when we follow the scale downward, and it can also be demonstrated in any number of songs or melodies. Each time, there is an intuitive feeling that the musical motion has to find a natural point of rest and a sense that the central tone literally attracts the melodic line toward itself. This symbolic attraction is also a driving force behind the progress of any melody. "All music is nothing more than a succession of impulses that converge towards a definite point of repose," wrote Igor Stravinsky.

How strong the attraction of the central tone is depends on the type of music we are dealing with. In homophonic Western music, which started to become the dominant type of music in Europe around the beginning of the seventeenth century and continued to be the most important form of music during the eighteenth and nineteenth centuries, the attraction of the central tone is very strong; it is perhaps the strongest type of attraction to the central tone that there is in any type of music. The central tone in homophonic music is known as the "tonic." Homophonic music is a derivative of polyphony, but the two types of music are, in many respects, quite different. The difference lies first of all in the fact that the basic structure of polyphonic music consists of simultaneously progressing melodies, while homophonic music consists of simultaneously sounding tones. After polyphony and actually as a consequence of polyphony, Western music became homophonic, or harmonic chord music.

There were many historical reasons for this development. By the end of the sixteenth century, polyphonic music became so complex that it could no longer endure further refinements and had really exhausted itself. Instrumental music had become increasingly important, and with it came new theoretical and practical problems concerning tuning since different instruments had to be made to sound well together. Furthermore, the desire to imitate the music of ancient Greece had led to the appearance of singing with accompaniment and thus again to the problem of different tones sounding together.* All these changes were leading to a new perception of music, away from the perception of simultaneously sounding melodies toward a perception of the effects of simultaneously sounding individual tones and so toward the eventual appearance of chords.

* This development originated in the mistaken belief that the Greeks knew about the rules of simultaneously sounding tones. They actually knew nothing about them, and their music, like that of all other civilizations except the late Western European, never used essentially different tones simultaneously in an organized manner. For European ears, however, whose perception was already amply educated by polyphony, simultaneously sounding tones did not sound alien at all. And therefore the acceptance of harmony-based chordal music was a relatively smooth process.

The new perception first evolved in Italy. Among its first theorists and advocates was Vincenzo Galilei, Galileo's father. The new style soon found its first great composer: Claudio Monteverdi. At this point, a new music theory also had to evolve. This theory was to concentrate on tone relations, on the problems of uniform tuning, all of which required the rationalization of scales and intervals on a mathematical basis. The new theory was summarized in several works, of which Jean-Phillipe Rameau's became the most famous. In these works, the basic rules governing the relations between chords were established and became known as the theory of harmony. From this time on, the intrinsic symbolic motion in music was represented largely by the motion of chords. This motion was also endowed with a clear way of signifying musical rest.

In classical Western harmony, there is essentially one chord which characterizes rest and it is the basic triad: the root and the third and fifth above it. If the first third is major and the second is minor, we have the major triad; if the first is a minor and the second is major, we have a minor triad (e.g., the major triad on "do" is "do-mi-so"). It is always one of these two chords which represents rest in Western harmony and toward which the other chords tend to gravitate. These basic chords are called consonances, the others are called dissonances. Despite its connotations in everyday usage, the word dissonance in music theory does not mean something unpleasant to hear. It means instead that such a chord cannot be used to finish a composition or even to finish a musical phrase—or, to put this differently, dissonances cannot represent rest in classical harmony.

Another important way to characterize consonant and dissonant chords is to say that dissonant chords represent tension, instability, or transition, while consonances represent resolution, finality, and rest. While in actual music, the loyalty to a central tone always seems to have existed, what exactly constitutes consonances and dissonances is a function of the musical language of a given society and period.

While the contrast between musical motion and musical rest, between tension and resolution exists in every kind of music, this contrast is most marked in harmonic chord music. It is more marked there than, for example, in polyphonic music or monophonic music because the "attraction" of the tonic-based consonant chord can be made very much stronger than the attraction of a single-note tonic. Obviously, in a chord, one feels (assuming again that the listener's ear was educated in this tradition) the attraction from several directions and from different components simultaneously. From immediately below, as in the step "ti-do" and from above ("re-do") and from many other intervals. Hearing several such steps simultaneously causes a strong demand for the resolution of tension, for arrival at the point of rest. This is why in the Western tradition, the dominant ("so-ti-re" in the above scale) or the dominant seventh chords ("so-ti-re-la"), for example, both create a strong anticipation of a resolution through the tonic chord. Some beautiful examples of resolutions are the harmonic cadences, i.e., the endings of musical phrases or movements or of the compositions themselves, in works of the classical period—in, for instance, the symphonies of Haydn, Mozart, and Beethoven.

The conclusion to be drawn is obvious: that the intuitive image of symbolic musical motion in classical Western music was perfectly compatible with the fundamental notions of motion in classical physics. All real motion in classical physics, as we recall, was imagined to be motion relative to an absolute rest, be that represented by Newton's "immovable center" of the universe or, later, by the ether. Every kind of physical motion, furthermore, took place in and could be regulated by a uniform, absolute metric time. The motion of bodies, as we also know, was always thought and perceived to be continuous, and it obeyed strictly causal laws.

Musical motion in the era of classical time and space could almost be described in the same terms. The symbolic motion of "sounding forms" also took place against that "immovable center" of the musical universe represented by the tonic chord, the symbol of absolute rest. The underlying progression of time, the background to musical motion, was also exactly measured metric time, manifested by the musical meters. The motion of chords was continuously unfolding, and obeyed logical and causal laws.

As in physics and in the visual arts, the beginning of the twentieth century saw a collapse in music of classical space-time concepts. In this case, what gave way were the notions of classical time, and with them went a world of continuous motion and absolute rest, of smoothness and predictability. And, just as in physics and in the visual arts, this collapse was not caused by sudden or unexpected events. The undermining of classical space and time had actually begun in the nineteenth century in all fields. Maxwell's theory of electrodynamics was, after all, the first theory which obeyed Einstein's special relativity, even if Maxwell never knew it. And it was Paul Cézanne who first suggested the abandonment of the Renaissance traditions in painting and tried to decompose visual objects into simple geometric forms, even as he continued to paint realistic paintings. And, similarly, it was during the nineteenth century that the role of the tonal center in musical works began to weaken.

It was the so-called romantic composers of the last century who extended the rules of harmony (in music, at least in the centuries between 1600 and 1900, the composer innovated and if the innovation turned out to be valuable, it became part of the theory), modulated into distant keys, and steadily diminished the organizing role of the tonic chord. By the end of the last century, such composers as Richard Wagner and Claude Debussy had come quite close to a practical rejection of tonality. But it was again in the twentieth century, in the compositions as well as in the theoretical writings of Arnold Schoenberg, that virtually all the traditions of classical music were consciously overturned. Beginning in 1908, Schoenberg began to advocate the development of a new musical language in which the tones would be organized through principles other than tonality. His method later became known as "atonality."

The significance of this development is that musical motion in atonal compositions no longer took place against a tonal center, could no longer be perceived as progressing against a symbolic rest. When in his later works Schoenberg advocated the equivalence of all tones and intervals, abolishing all distinctions between consonances and dissonances, he was making musical motion fully "relative." The distinction between motion and rest—and thus absolute motion itself—was abolished in music as well.

But there was actually more than one way in which twentieth-century music abandoned the images of smooth, continuously passing classical time. The break with the old perception of time, in other words, is apparent even if we disregard Schoenberg and his school. The use of polyrhythms, for example, a technique used by the greatest composers of our century, Béla Bartók and Igor Stravinsky most prominently among them, is an example of these radical changes. Polyrhythm is a musically powerful and effective method using strongly contrasting rhythms and even contradictory musical meters simultaneously. The image of temporal motion in such works is not an image of uniformly passing, smooth, metric time—indeed, it is far removed from it. Interestingly, this expressive technique was virtually nonexistent during the period when Europe held to the classical notions of time—i.e., from about 1400 to 1900. It had been used, in contrast, before 1400 and was found again after 1900 in Europe and has been utilized continuously in non-Western civilizations in Africa and Asia.

An equally interesting and definitely nonclassical form is *polytonality*. Bartók seems to have been the first to use it, and the method was developed further by Darius Milhaud in France and Sergey Prokofiev in Russia. Instead of using a single tonal center or a single system of rest, the composer would use two or three such systems simultaneously. The musical effects are different from those in an easily understood tonal piece but are not as startling as in music without any tonal center at all.

The ideas of Schoenberg were, however, by far the most influential in shaping the musical language of the twentieth century. The application of his principles made atonal music very different from every other kind of music, classical, popular, folk, or art, Western or non-Western. As there is in this music no background at rest, one finds it difficult to perceive musical motion at all. For this reason, perhaps, many people find it difficult to concede that the succession of tones and chords in atonal music can be characterized as music at all. If a musical motion is without tonality, it appears to fail to correspond to any intuitive mental model of motion we have in our nervous systems. We instinctively perceive such musical motion to be arbitrary, discontinuous, jerky, and difficult to make sense of. The strange laws it obeys may be formally or intellectually understood, but the end product does not correspond to our inborn intuitions or "fit" our normal perceptions.

There were few composers in the second half of this century whose music was not influenced by Schoenberg's ideas. At the same time, however, the influence of these very same ideas remains relatively minor in the concert halls. Music without a tonal center does not seem to have become as accepted by the general educated public as, say, paintings without literal content. There are several possible explanations for this fact. It is possible that this type of music simply does not make sense to most listeners even after hearing it many times. It may simply be much too abstract. It is also possible that this type of music is still waiting for some genius-composer who could make it more meaningful.* Time will tell.

* Or one can argue more maliciously that abstract paintings were found to be less objectionable by the general public because it was always easier to walk by a painting making approving noises than to sit in a concert hall and actually listen to difficult music for an extended period of time.

10
The Present
(Nothing but Spacetime?)

The last rung in the ladder of knowledge is
the most abstract and subtle of all.
Saadia Gaon

To unify, to integrate diverse pieces of information, is what brains are for. Without this particular mental ability, we would live in a world of disconnected impressions and reflexlike responses, a world without extended space and time. Like the history of the brain, the history of physics is also a history of unifications. Newton unified the mechanics of the sky with that of objects on earth. Maxwell unified electricity, magnetism, and light. Einstein unified space, time, matter, and motion. Recent studies aim at unifying all the known interactions of nature as well as unifying the laws of the very small with the laws of the large-scale universe. A sketch of some new ideas about these matters is given here. If these ideas are correct, they will again change what we think about the nature and role of spacetime.

To start with, there now exists rather convincing evidence that the phenomenon we call the expansion of the universe—i.e., the fact that all galaxies in the universe are receding from each other—is a consequence of a single event: a primordial explosion, commonly known as "the big bang," which started the expansion of three-dimensional space and matter. The most important evidence turned up in the following way. Since Hubble's discovery in 1924 that the universe is expanding, astronomers have been able to determine with increasing accuracy the law which describes how fast the galaxies were receding from each other. Knowing this data and assuming that the process of expansion did indeed start with an explosion and that the law which determines its rate has not changed, it became possible to calculate when the explosion occurred. This calculation yielded the approximate value of about fifteen billion years, the figure that is usually referred to as "the age of the universe."

In speaking about the primordial explosion, it is plausible to assume that, among other characteristics, the state of matter at the time of that event must have been unimaginably dense and hot. Dense because a relatively small space had to contain all the matter now in the universe, and hot because it had to supply all the energy now existing in the universe. The cosmic expansion started, therefore, with an explosion of hot, dense matter. To learn what happened after the explosion, it is helpful to recall a long-established law of nature which says that when a hot gas expands in space, it always cools off. Consequently, as the matter in the universe expanded, its temperature must have dropped.

Among the various constituents of the present universe, photons roam cosmic space relatively freely, and they are also relatively easy to detect. They constitute a "gas" which started to crisscross the universe soon after the primordial explosion and at a time when the universe was still very hot. Using normal physics one can calculate that in about fifteen billion years the photon gas should have cooled to a temperature of about a few degrees Kelvin, i.e., a few degrees above absolute zero.

In 1964, two scientists at Bell Telephone Laboratories in New Jersey, Arno Penzias and Robert Wilson, discovered in the course of preparing a completely unrelated work in radio astronomy that the universe was literally bathed in radiation at a temperature of 3 K.* This radiation comes from all directions, permeates space everywhere, and belongs to that part of the electromagnetic radi-

* The radiation, to be more exact, is the same as that emitted by a blackbody at a temperature of 3 K. In our case, the blackbody is the entire universe.

ation spectrum that we call microwaves. The unexpected discovery of radiation showing just the required properties made a strong case for the theory that the universe indeed started about fifteen billion years ago with what we roughly imagine as an explosion of hot, dense matter.*

Having established this much, we can also see that we should be able to learn about the history of the universe and about the processes by which it reached its present state if we could learn how matter behaves in the kind of extreme conditions which prevailed during and right after the primordial explosion. If we could learn, in other words, about the behavior and mutual interaction of very energetic particles. Such particles, however, cannot be found on earth or in the solar system under normal conditions. To study them we literally have to create them, and for that we need machines to accelerate particles, thus endowing them with very high energies. Then we can investigate how such particles interact with each other. In these very powerful accelerators, we can study those processes which were important when the universe was very young.

It is from the data supplied by such experiments that physicists hope to reconstruct with some confidence the early history of the newborn universe: the time sequence of those primordial events which determined the way the universe looks now. The mere collecting of these data would not be sufficient, however. Theories which connect and make sense of them are also necessary. Among the theoretical tools which help us to unravel the history of the universe are the equations of the general theory of relativity. These equations give us, among other things, a mathematical connection between the age of the universe and the density of matter it contains. Since it is the density and temperature of

* The first theory proposing the primordial expansion of hot matter was suggested in 1948 in a research paper ostensibly authored by R. Alpher, H. Bethe, and G. Gamow. The second author, Hans Bethe, an eminent physicist now retired from Cornell University, had actually done no work on the paper. The paper was written entirely by the Russian-American physicist G. Gamow and his student R. Alpher. Gamow, however, was a famous practical joker and had a fine sense of symbolism. He argued that a paper dealing with the beginning of all chemical elements should have three authors with names resembling the first three letters of the Greek alphabet. To achieve this, he simply put Bethe's name on the paper.

matter and radiation which largely determine what type of particles can exist permanently and what sorts of physical processes take place in a system, the above connection enables one to reconstruct the sequence of those early events. We have learned, to give an interesting example, that up until a few hundred thousand years from its beginning had passed the energy of the universe was concentrated not in ordinary matter as today but in radiation. Light, that ultimate symbol of "lightness," was therefore during the first three or so hundred thousand years quite dense everywhere in the universe. Much denser than, say, heavy oil.

The mental reconstruction—based on the big-bang theory—of the steps through which the universe evolved worked very satisfactorily in the sense that it made understandable some of the most crucial observed features of the universe. It explained, as we know, the fact of the expansion itself, as well as the fact that the universe is bathing in the cold afterglow of creation: the microwave radiation. In addition the big bang theory also explained the relative abundance of the lightest elements: it can be calculated, for example, how much more hydrogen than helium exists in the universe and the numbers fit well with observations. The time-range in which the big-bang theory seemed to work was also quite amazing. The theory was able to reconstruct the history of the universe from the time the universe was less than a second old to the present; i.e., about fifteen billion years later. When in 1976 Steven Weinberg wrote a popular book about the first three minutes of the universe on the basis of the big-bang theory he had to leave only the first one-hundredth second unexplored.

No theory can work so well accidentally. Its very success showed quite convincingly that the big-bang theory was essentially on the right track. In spite of all this, however, all was not well with this theory as it was formulated before 1981. While it explained a number of important observed phenomena there were other also rather fundamental facts which the theory could not explain at all. Without going into details I shall mention here one single example*: the problem of the *uniformity* of the universe. As

* This topic is much too big and complex for the thumbnail sketch I can give in this short closing review. Also, our concern in this chapter continues to be the nature of space and time and not cosmological theories per se. About the latter there exist several recent nontechnical books worth reading.

far as it is possible to know, the distribution of matter as well as radiation is exceedingly uniform in the visible universe. Galaxies, and the clusters made up of many galaxies, are distributed essentially similarly, near and far, left and right, up and down. If, initially, conditions were significantly different at different parts or directions in the universe this could not be so. Also such differences would surely have shown up in the distribution of the microwave radiation. In fact all the properties of the radiation are the same from all directions to a very high degree of accuracy. One can conclude from all this that no matter where you were situated in the universe you would always observe the same overall characteristics as you do from the earth. There is, in other words, a kind of large-scale uniformity in the universe as if the universe was molded everywhere by very similar processes.

It is this large-scale uniformity which is not possible to understand from the premises of the big-bang theory for the following reasons. One of the basic assumptions of this theory is that the law describing the rate of the expansion has not changed since the beginning of the universe. It follows from this fact that there are regions in the universe which could never have been in any causal contact with one another. The following example explains this: Consider two distant galaxies which we now observe in different directions. The distance between them could be so great that the time it would take light to reach one from the other is definitely larger than the age of the universe. Since nothing could ever have traveled faster than light our two galaxies could never have been in any causal contact; i.e., what happened to one could never have influenced the other. There are in fact a very large number of regions in the observable universe which, for the same reason, could never have been in any kind of causal contact. How was it possible then that all these regions emerged from the tremendous chaos and turbulence of the big bang exhibiting the same overall order?

To overcome this and a number of other similarly vexing problems, Alan Guth (then of Stanford University) proposed in 1981 a new and more detailed model to describe how the universe came into existence. His ideas were improved upon in 1982 by Andre Linde of Moscow University and by Andreas Albrecht and Paul Steinhardt of the University of Pennsylvania. This new model is based on current ideas in elementary particle theory and what this theory says about the behavior of matter in circumstances which probably existed at the very beginning of the universe. One of the striking features of the proposed new model is that it suggests that the properties of the present universe were already determined at times unimaginably shorter than one second. According to this theory the universe started from a state in which its spatial extension was a tiny fraction of the size of a single elementary particle and which state also contained very little matter. The universe in its initial state, in other words, was essentially vacuum and occupied a fantastically small region of space. The word "vacuum" means—here and elsewhere in this discussion—Dirac's vacuum: nothing at all actually, yet potentially a huge reserve of matter and all properties of matter. Now one of the new insights gained in the recent advances of elementary particle theory was that the vacuum could have existed in different "states," meaning that its energy content could have had different values. The vacuum, in other words, much like an atom, could exist at various energy states. This turned out to be a crucially important fact.

It is next assumed, and current elementary particle theories make this assumption quite plausible, that very shortly* after its beginning, the universe entered into a specific vacuum state which, however, was not the vacuum's lowest allowed energy state. This state is called the "false vacuum." Calculations show that the false vacuum has extraordinary properties. Among other things, it exerts an enormous effect on space itself. This effect is usually called, for lack of a better word, "repulsion." But this is not repulsion in the ordinary sense of this word as in the sense of

* "Very shortly" seems to be a vast understatement. It means more precisely about 10^{-34} seconds after the beginning of the process whose endproduct was our universe. 10^{-34} seconds is an incredibly small time interval. The age of the universe is about 10^{17} second. Imagine now the time interval which goes into one second as many times as seconds go into the age of the universe; i.e., 10^{-17} seconds. Call this very small time interval, say, a *whit*. Divide now a whit again into 10^{17} equal parts and what you get is the incredibly small time interval of 10^{-34} seconds. All this sounds quite fantastic but in this case the extreme numbers of cosmology and particle physics happen to combine to give such incredibly small time intervals.

two like electric charges repelling each other *in space*. The repulsion in the false vacuum acts *on space itself* and forces it to expand at a fantastic rate. Incredibly faster than the rate one would expect from observing the present expansion of the universe. It is incidentally because of this stupendous expansion rate that this early phase of the evolution of the universe was dubbed as "inflationary."

The inflationary period had two important characteristics. One is that the expansion was so fast during the period of expansion that the "edge of space" would have moved away from an observer (if we imagine one to exist in this fantastic setting) much faster than the speed of light. This does not contradict the laws of physics since we do not associate any kind of matter or energy with this speed. It is empty space itself which expands its limits via such a fast process. The other also rather startling property of the inflationary period is that according to calculations the false vacuum actually *increased* its energy content during the expansion. This is very strange, because no normal physical system can be imagined which would do the same. Normally an expansion would diminish the system's energy, not increase it. The false vacuum is a strange physical state which probably no longer exists anywhere in the universe, which came into existence when the universe was about 10^{-34} seconds old and lasted for about 10^{-30} seconds altogether. But during this very short time the inflationary process tremendously increased both the size of the universe and its energy content.

The end of the inflationary period comes when the false vacuum spontaneously decays into the lower energy state called the "true vacuum," which is essentially the vacuum of the present universe. In other words, in the same manner as when an electron decays from a higher energy to a lower energy state a photon is created, when the vacuum decayed from its "false" state to its "true" state the universe was created. The energy the false vacuum acquired through its expansion became liberated at this decay and it exists now as the total mass and energy of the entire universe. The decay of the false vacuum in other words was the act from which the universe—in its present form—originated, and this decay, the transition from the false vacuum state to the true one, was what we called earlier the big bang. But instead of considering it as the absolute beginning we now think that the big bang came after the first 10^{-30} or so seconds of the universe during which time virtually all the important properties of our universe had already been determined. The decay of the pre–big bang false vacuum supplied all forms of matter and energy within the universe including the kinetic energy of matter which is manifested in the expansion of our universe. This expansion, according to this theory is the much slowed down continuation of the primordial inflation. Interestingly, it also follows from the inflationary theory that the region we call the "universe" is probably just a fraction of the entire system which came into existence in the same process.

Since the inflationary theory incorporates the original big-bang model, the two theories become identical within a very short time. It is clear then that the inflationary theory explains everything the big-bang theory can explain. This theory is, in other words, at least as good a theory as the original big bang. But the inflationary theory is also able to account for phenomena which were inexplicable by the original theory. Take the example mentioned above: the problem of the uniformity of the universe. Since, according to the inflation theory, the universe started out from an exceedingly small spatial region, and it expanded far faster than anybody could have imagined on the basis of the big-bang theory, all parts of the present universe could have been in causal contact at the beginning, and could have therefore acquired essentially similar properties.

In light of this theory, it is also tempting to think about the "cause" of the whole process. Why did it happen? Attempts to answer such questions are highly speculative and will probably remain so for a long time to come. One possibility is that the creation of the universe was a statistical, random event like the decay of a radioactive nucleus or the transition of an electron from one state to another. It had no "cause" in the usual sense of this word. If this was indeed so then the words of Edward Tryon of the City University in New York described creation quite well. Tyron remarked while discussing another "universe from vacuum" idea: "In answer to the question of why it happened, I offer the modest proposal that our universe is simply one of those things which happen from time to time." Maybe so. But do not place too high a bet on it. It is conceivable just as well that the universe is an event which could happen once only. So, as we see, the ancient problem of periodic vs. linear time is still with us.

An important feature of contemporary cosmology is that the universe and most likely space and time with it came into existence through physical processes which, while unobservable in themselves, were extrapolated in a rational manner from well known observable ones. It is all the more intriguing then that the whole system of the world—space, time, matter—all seem to have sprung into existence from nothing. "I have often heard it said that there is no such thing as free lunch. It now appears possible that the universe itself is a free lunch," said Alan Guth, the creator of the inflationary theory. The universe indeed seems to have come from the vacuum; i.e., from nothing. In some sense the primordial vacuum is nothing; it may be defined as nothing: No matter, no space, no time. On the other hand, it is also a seemingly inexhaustible fountainhead and reservoir of everything which exists—just the opposite of nothing. It would perhaps be best to forget the word "nothing." It may be one of those many human symbols which once created, continues to haunt us. There may be no nothing at all.

In addition to helping us toward understanding the birth pangs of the universe, contemporary theories of elementary particles are shedding new light on the very nature of time and space. This fascinating research began in the 1930s when it became known that there are four essentially different forces or interactions in nature. Everything that happens in the universe is a result of the action of one or more of these forces. Gravitation is the force we have known about for the longest time, but is also the one which is no longer thought of as a force or interaction, having instead been transformed into the structure of spacetime in Einstein's hands. The second force or interaction is electromagnetism. This is the interaction which creates light, drives all our electrical machinery, and keeps atoms as well as our own bodies together, among other things. Both gravitation and electromagnetism have "long" ranges—i.e., their actions extend to perceivable distances (to infinity, mathematically speaking)—and therefore both were "discovered" long ago. The other two interactions, on the other hand, those called the strong and the weak interaction respectively have short ranges (less than a trillionth of a centimeter) and their effects are not directly observable. (Those forces which keep the atomic nucleus together are manifestations of the strong interaction, while the weak force is responsible for some forms of radioactivity and determines the stability of atomic nuclei in the universe.)

Perhaps the most important result in physics during the last two decades was the discovery that two of these forces are, in fact, different manifestations of a single interaction. A highly imaginative theory was worked out in the late sixties which unified the electromagnetic and weak interactions.* In 1982, experiments with the huge accelerators at the European Center for Particle Research in Geneva confirmed all the important predictions of this new theory, among them the existence of a neutral particle (called the "Z boson"), a quantum mechanical heavy cousin of the photon, sometimes dubbed "heavy light." This unified interaction is called by the neither melodic nor inventive name *electroweak*.

The success of the electroweak theory reawakened an old idea, the idea of further unification, further simplification, in the minds of many physicists. Einstein spent the last decades of his life trying to construct a unified theory of the two long-range forces, gravitation and electromagnetism, in the framework of nonquantum physics. None of his attempts turned out to be relevant. It is likely that the unification of the three (gravitation, strong, electroweak) known interactions must be achieved through the ideas of quantum theory. This makes the problem more challenging, as well as more inviting. It seems, in fact, that with the attempt to unify all the forces or interactions of nature, physics has embarked on the most ambitious project in its entire history.

The questions which are now being asked are these: Is it possible that all the varied interactions in nature are just different manifestations of a single fundamental interaction, and is it possible that a single force governs and determines everything that happens in the universe? The implications of an affirmative an-

* The final, testable form of this theory was constructed independently by Steven Weinberg, then of Harvard University, and Abdus Salam of Imperial College, London.

swer are staggering. All that exists in the universe results from interactions. Consequently, if a single interaction is responsible for everything in the universe, then all particles must be just different states of *one single particle;* all matter in the universe could be reduced to a single constituent. The fabric of the cosmos would then be shown to be the ultimate in simplicity.

The theories aiming at finding the key to this apparently final unification are called supersymmetric theories (it will soon become apparent what the significance of symmetry is in this context). A successful supersymmetric theory would be the ultimate in generality, in simplicity, and in explanatory power. It is also quite possible, as we shall see below, that such a theory would say totally new and unexpected things about the very nature of spacetime.

The question which arises immediately in connection with a unifying theory is this: If the basic design of the universe is indeed the ultimate in simplicity, then what makes the actual appearance of the universe relatively complicated? It is now generally believed that the answer to this question lies in a natural process called (mainly for historic reasons) "spontaneous symmetry breaking." The credibility of this assumption has been greatly enhanced by the success of the electroweak theory. In this theory it is the process of spontaneous symmetry breaking which explains how two seemingly very different interactions can have a common origin and, in fact, be the same interaction at high energies.

The idea of symmetry is fundamental in modern physics and will probably continue to be so in future theories of space and time as well. What do we mean by "symmetry"? The meaning emerges when we note that real or imaginary objects can retain certain properties after being manipulated. We call a white sphere a highly symmetric object because we can rotate it around its center any way we like and it will always look the same, or in other words, will retain all its visual properties. We can rotate a uniformly black cylinder around its horizontal axis and it will still look the same. We can look at a Gothic arch in a mirror and it will look like the original. These objects exhibit symmetries which are geometrical: The first is called spherical and the second cylindrical, while the third is the left-right symmetry. But it is not only objects which can be symmetric. The sentence "You can cage a swallow, can't you, but you can't swallow a cage, can you?"* also shows a definite symmetry; it sounds the same whether you read the sequence of the words forward or backward. Composers throughout history have played with melodies which remained unchanged when sung or played backwards.

Some manipulations can "break" the symmetry. If we paint the upper, or "northern," half of our white sphere black, then only rotations around the north-south axis retain the sphere's visual appearance. The spherical symmetry is broken and is replaced by a cylindrical symmetry. If we paint random figures on our sphere, the symmetry may be completely broken. We may see a different sight at each rotation. The essential visual symmetry of the spherical shape is, in this case, concealed by the added random figures.

It can happen that material systems exhibit a symmetry in a high-energy state but that this symmetry is broken and becomes hidden when the system loses some of its energy. The magnetic behavior of iron is a standard example. The molecules of this material always behave like tiny magnets. As long as the iron is hot, the energy of the motion of the molecules overwhelms the small energies of their magnetic interaction. Consequently, the north-south axes of the tiny magnets are randomly oriented and there is no preferred direction in the system. No matter which direction an observer positioned inside the metal were to look, he or she would always see the same picture: randomly oriented magnetic molecules. The system exhibits perfect symmetry, which is as it should be because the basic mathematical laws which describe such a system happen to display such a symmetry. In cold iron, there is not much molecular motion, and the interaction between the individual magnets is therefore significant. The normal physical state of cold magnetic iron is determined by the fact that, like every physical system, the magnetic molecules also tend to be in the state which has the lowest energy. It so happens that for magnetic iron the lowest energy state is one where the molecules all line up in one direction. The symmetry of the hot state does not exist in this state. An observer inside the magnet will no longer see the same sight in all directions. This so-called spontaneous breaking of the symmetry now hides the essential symme-

* Borrowed, of course, from Martin Gardner.

try of the basic laws of the system. The symmetry exists, but it is hidden from view and can be reestablished only if one imparts energy to the system by heating it up.

The modern unification theories of physics assume that something similar once happened on the largest scale. They assume that perfect abstract symmetries characterize the ultimate laws of the universe. These symmetries were all present at the beginning of the cosmic evolution when all matter was very hot and energetic. As the universe expanded and cooled, some of the symmetries were broken and became hidden. In order to see them again, we have to study those processes which involve very high energies. By studying very energetic processes in large accelerators, we hope to catch a glimpse of some of the symmetries which existed at the very beginning of the universe.

There are a great many symmetries associated with the laws of physics. Some are easy to visualize, like the above mentioned spherical symmetry or left-right symmetry. Others are somewhat more abstract. The laws of motion in classical mechanics, for example, would not change if the direction of time were reversed. If you looked at a movie of the motion of the planets around the sun, a knowledge of the laws of planetary motion would not enable you to decide whether the movie was projected forward or in reverse. That the two processes are identical is a manifestation of the temporal symmetry (a symmetry under the mental reversal of the direction of time) built into the classical laws of motion.

The laws of physics often possess symmetries which do not involve space or time. These are abstract or "internal" symmetries, which cannot ordinarily be visualized at all. Nevertheless, they are very important to an understanding of nature's laws. For easier comprehension and for better visualization, these symmetries are often expressed in abstract symbolic spaces. In elementary particle physics, to give an example, the so-called isotopic spin space plays an important role. The "isotopic spin space" has well-defined mathematical properties, its geometry is easy to comprehend, but it has no connection whatever to ordinary space. On an "axis" in isotopic spin space, "up" represents the possibility that a particle may be a proton, "down" that it may be a neutron. The other two axes are imagined to be mathematically "perpendicular" to the first one. Such a space is, of course, a purely mental construction. But it is also a framework for an often useful code, for a specific symbolic representation of certain aspects of reality. Its usefulness lies in the fact that geometric properties of this space can be translated into entirely nongeometrical features of the real world. Thus, the geometric statement which says that the laws of the strong interactions are symmetric under rotations in their abstract space means, when translated into physical facts, that strong (nuclear) interactions are not influenced by electric charges. The use of abstract spaces makes the mathematics simpler to handle and gives insights as to the nature of the phenomena. Such spaces have become indispensable in research.

There are all sorts of interesting symmetries in nature. Some are more basic than others. The hexagonal symmetry exhibited by a benzene molecule is important theoretically and practically, but this symmetry is unlikely to represent such a basic feature of the evolution of the universe as, say, the directional symmetry of the expansion of the universe (i.e., the fact that the expansion is the same in all directions).

At present, researchers are mainly interested in apparently fundamental, abstract mathematical symmetries known as "gauge symmetries." The name merely reflects the historic fact that the first theory of gauge symmetries was concerned with symmetries manifested under the "regauging" of distances and time intervals. In vastly generalized forms, these symmetries play a crucial role in electroweak theory as well as in current theories of strong interactions. What interests us here is that all four forces of nature display gauge-type symmetries and that many physicists think that this fact may be a clue to some underlying unity of the forces of nature.

In physical theories, abstract symmetries in general and gauge symmetries in particular give rise to phenomena that we can actually observe and measure. Yet these symmetries themselves do not refer to the physical world, they are not symmetries of anything real. They are purely mathematical. The question then arises: What is it in nature which corresponds to these abstract symmetries? How does nature realize the properties which we discover in abstract mathematical forms? Are there perhaps some hidden properties of reality, not as yet discovered, which actually give rise to the gauge symmetries? At the present time, many physicists think that these symmetries are not only real but are manifestations of *actual spatial symmetries* which exist in *higher spatial dimensions.*

The thought of higher spatial dimensions conjures up pictures from the modern mythologies of science fiction. Yet there is sober reasoning behind the idea of and the search for hidden dimensions of space. These efforts represent an actively pursued direction in the quest for a unified theory.

The beginning of the twentieth century saw a unification of space with time in Einstein's special theory of relativity. Ten years later, further unification was accomplished in the general theory, which gave a unified picture of spacetime and gravitation. Very soon after the appearance of this theory, people became interested in what looked then like the next logical step: the unification of spacetime and gravitation with electromagnetism. Since gravitation and electromagnetism were the only known forces in nature at that time, their unification appeared to be a most important and logical step. Some even thought it to be the ultimate step. Since the general theory of relativity showed that gravitation was essentially a manifestation of the geometric structure of spacetime, the aim of those unified theories was to interpret electromagnetism in exactly the same way. A new geometry of spacetime was to be created which put gravitation and electromagnetism on the same footing, which treated both as manifestations of the structure of spacetime. As I mentioned earlier, Einstein himself devoted the last thirty years of his life to this task, but neither his nor others' attempts were successful then or later. And when the strong and weak interactions were discovered in the 1920s and 1930s, respectively, making it clear that there were other forces in nature besides gravitation and electromagnetism, the interest in the problem eventually died out.

Before all this happened however, in 1919 to be more exact, two theories were suggested which, while unsuccessful, nevertheless contained some profound ideas which influenced the development of physics half a century later. One was the unified theory of gravitation and electromagnetism propounded by the Swiss mathematician Hermann Weyl. The theory did not get anywhere, but it was the first which suggested and explored the use of gauge symmetries. The other was the work of a then young and unknown Polish mathematician, Theodore Kaluza.

Kaluza proved an interesting mathematical fact: If we assume that space has a *fourth* dimension, i.e., assume that spacetime has altogether *five dimensions*, then both gravitation and electromagnetism can be interpreted as being nothing but manifestations of the geometry of this new, enlarged spacetime. If, for example, you considered Einstein's theory in five dimensions, then you would find that a particle would still move on the shortest path between two points, but now in five-dimensional spacetime. You would then be able to calculate how this path would look to a human observer who lives in three-dimensional space and one-dimensional time. You would find that the path looks exactly like that of a particle moving under the combined influence of electromagnetism and gravitation. No wonder that Einstein as well as others found the theory highly interesting. All these results, however, turned out to be purely formal. One great practical drawback of Kaluza's theory was that it failed to answer one crucial question: If space is indeed four-dimensional, why can we perceive only three? Where is the fourth one? Because the theory gave no answer to this question, it was considered an interesting and ingenious mathematical exercise but not much more.*

Five years after the publication of Kaluza's theory, it was revived by Oscar Klein of Stockholm University. Klein applied Kaluza's idea to the then brand-new quantum mechanics. He was able to show that Kaluza's results can be extended to the atomic domain, and, equally important, he suggested a reason why we cannot see the fourth dimension of space. Klein calculated that the fourth spatial dimension was very small, that it does not extend like the other three to vast lengths but is "rolled up" to a minute extension, and this is the reason that we do not perceive it. This, at least, is a rational explanation and easy to understand. If you look at a fine thread or at a thin line on a paper, you see it as

* The first natural and spontaneous objection to such a theory, that we are completely unable to imagine a fourth spatial dimension, carries little weight. Experience has repeatedly taught us that nature does not care about what we can or cannot imagine. Mathematics works well in any number of dimensions, and that is all that is needed to make sense of them.

essentially one-dimensional. Only by using a magnifier will you reveal its hidden width, the second dimension. This dimension was not perceivable to you before because of its smallness. Klein calculated the extension of the fourth dimension of space on the basis of the quantum mechanics of the electron and found it to be about 10^{-32} centimeters. This is an incredibly small distance, very, very much smaller than, for example, the diameter of an atomic nucleus. No wonder that we cannot experience it and that our nervous system has no model for it. It seems astounding at first that such a small extension into the fifth dimension should be enough to produce all those large effects of electromagnetism. But we should remember that an extension into the fourth spatial dimension would exist at every point in three-dimensional space. Therefore, the totality of the extension into the fourth spatial dimension could be substantial.

But in spite of the interesting results of Klein's paper, the Kaluza-Klein theory, as it came to be called, still fizzled out relatively soon. It could not generate new investigations at the time; its proponents could not propose new experiments which would have tested the statements of the theory against the real world. Like many other well thought out, well-constructed physical theories, it fell by the wayside because it could not be related to new observations. It did not explain previously unexplainable facts.

During the last decade, the basic ideas of this theory have been resurrected and extended. We are no longer interested only in the unification of gravitational and electromagnetic interactions. When we speak about unification now, we are interested in a unified theory of all the interactions of nature. As I mentioned earlier, a common property of all interactions is that they exhibit the abstract mathematical symmetries which go under the name of gauge symmetries. A working hypothesis, which is now being actively explored, is that those aspects of reality which manifest themselves in these abstract symmetries are actually *geometrical symmetries of several hidden spatial dimensions.* In other words, the gauge symmetries of abstract mental space actually express the symmetries of real physical space in dimensions which we cannot directly perceive.

It stands to reason that the one extra spatial dimension of the Kaluza-Klein theory which unified gravitation with electromagnetism is no longer sufficient. To accommodate all the gauge symmetries of gravitational, strong, and electroweak interactions, more dimensions are needed. How many? This depends on the particular version of the basic idea, on the particular mathematical theory. At present, some researchers are working on the assumption that there are seven extra spatial dimensions. This would mean that the real world actually has eleven dimensions, four of them extensive (the three spatial and one time dimension of our normal world), and seven which are curled or rolled up and have minute extensions only.

Why this division of four plus seven or some other number of dimensions for that matter? There is perhaps some deep reason for it which we do not now understand. But it may also have been an accident. The universe could have started out with ten spatial dimensions and one temporal dimension, all on an equal footing. Some primordial turbulence (inflation perhaps) then could have caused three spatial dimensions to expand, and left seven dimensions unchanged or even shrinking. Perhaps shrinking at the same rate as the three expanded. One day we may learn all the details.

These theories are all in their infancies, far from being proved, and there are several competing theories aiming towards the same goal. Some explore similar ideas, others work along entirely different lines. Time will tell which, if any, will turn out to be the real thing. But the possibility that all the forces which act in the universe could turn out to be manifestations of the structure of spacetime of whatever dimensionality boggles the mind. For *these forces include everything which exists in the universe.* Every particle, every form of energy, every piece of reality, is a product of these forces. There is nothing else in the universe. The general theory of relativity established that it is the structure of four-dimensional spacetime which manifests itself as gravitation. With similar logic, a spacetime theory of all interactions would establish decisively that there exists nothing in the world but structured spacetime. Certain structures would manifest themselves as galaxies, others as quarks or black holes, flowers or bacteria, yet others would be you or me. If this is really so, then it would surely have been an event of cosmic significance when spacetime became conscious of itself. We would then know that this happened at least once, on a small planet circling an average star.

Bibliography

Abraham, G. *The Concise Oxford History of Music.* London: Oxford University Press, 1979.

Apel, Willi. *Harvard Dictionary of Music,* 2nd ed. Cambridge, Mass.: Harvard University Press, 1972.

Arnheim, R. *Art and Visual Perception.* Berkeley: University of California Press, 1969.

Arons, A. B., and Peppard, M. B. "The English Translation of Einstein's Photon Paper." *American Journal of Physics, 3:3* (1965). p. 367.

Asimov, I. *The Universe from Flat Earth to Black Holes and Beyond.* New York: Walker and Co., 1980.

Bartschi, Willy A. *Linear Perspective.* New York: Van Nostrand Reinhold, 1976.

Bedini, Silvio A. "The Scent of Time." Transcript of the American Philosophical Society, 5:3, (1963). p. 5.

Bergmann, Peter G. *The Riddle of Gravitation.* New York: Charles Scribner's Sons, 1968.

Bernstein, J. *Einstein.* New York: Viking Press, 1973.

Blakemore, C., and Cooper, G. F. "Development of the Brain Depends on the Visual Environment." *Nature, 228,* (1970). p. 477.

Bloom, A., trans. *Plato: The Republic.* New York: Basic Books, 1968.

Bondi, H. *Relativity and Common Sense.* New York: Doubleday Anchor Books, 1964.

Born, Max. *Einstein's Theory of Relativity.* New York: Dover Publications, 1964.

Boulding, Kenneth E. *The Image.* Ann Arbor: University of Michigan Press, 1956.

Boyer, Carl B. *A History of Mathematics.* New York: John Wiley & Sons, 1968.

Brady, John. *Biological Clocks.* London: Edward Arnold, 1979.

Brady, John, ed. *Biological Timekeeping.* Cambridge: Cambridge University Press, 1982.

Bukofzer, M. *Studies in Medieval and Rennaisance Music.* New York: W. W. Norton, 1950.

Bunning, E. *The Physiological Clock.* New York: Springer-Verlag, 1973.

Butterfield, Herbert. *The Origins of Modern Science 1300–1800.* Toronto: Clarke, Irwin, 1957.

Caldwell, John. *Medieval Music.* Bloomington: Indiana University Press, 1978.

Campbell, Joseph. *Myths to Live By.* Toronto: Bantam Books, 1972.

Chandrasekhar, S. "Beauty and the Quest for Beauty in Science." *Physics Today,* July, 1979. p. 25.

Chew, G. F., Gell-Mann, M., and Rosenfeld, A. H. "Strongly Interacting Particles." *Scientific American* Vol 210. No. 2, p. 74, 1964.

Clagett, Marshall. *The Science of Mechanics in the Middle Ages.* Madison: University of Wisconsin Press, 1959.

Claudsley-Thompson, J. *Biological Clocks.* London: Weidenfeld and Nicolson, 1980.

Cold Spring Harbor Symposia. "Biological Clocks." Volume 25, 1960.

Copland, Aaron. *The New Music 1900–1960.* New York: W. W. Norton, 1968.

Copland, Aaron. *What to Listen for in Music.* New York: The New American Library, 1957.

Cornford, Francis M. *Plato's Cosmology: The 'Timaeus' of Plato Translated with a Running Commentary.* London: Routledge & Kegan Paul, 1948.

Cornford, Francis M., trans. *The Republic of Plato.* Oxford: Oxford University Press, 1968.

Craik, K. J. W. *The Nature of Explanation.* Cambridge: Cambridge University Press, 1952.

Davies, Paul. *The Forces of Nature.* Cambridge: Cambridge University Press, 1979.

Davies, Paul. *Other Worlds: Space, Superspace and the Quantum Universe.* London: J. M. Dent & Sons, 1980.

Davies, Paul. *The Runaway Universe.* Harmondsworth: Penguin Books, 1978.

Davies, Paul. *Space and Time in the Modern Universe.* Cambridge: Cambridge University Press, 1977.

Davies, Paul. *Superforce: The Search for a Grand Unified Theory of Nature.* New York: Simon & Schuster, 1984.

Davis, Philip J., and Hersh, Reuben. *The Mathematical Experience.* Boston: Houghton Mifflin, 1981.

De la Croix, Horst, and Tansey, Richard G. *Gardner's Art Through the Ages,* 5th ed. New York: Harcourt, Brace & World, 1970.

De Solla Price, Derek. *Science Since Babylon.* New Haven, Conn.: Yale University Press, 1961.

Dillard, Annie: *Living by Fiction.* New York: Harper & Row, 1983.

Dittner, L., trans. *Anonymous IV.* New York: Institute of Medieval Music, 1959.

Drake, Stillman. *Galileo at Work.* Chicago: University of Chicago Press, 1978.

Edgerton, S. Y. *The Renaissance Rediscovery of Linear Perspective.* New York: Basic Books, 1975.

Einstein, Alfred. *A Short History of Music.* New York: Random House, 1937.

Eliade, Mircea. *A History of Religious Ideas,* vol. 1. Chicago: University of Chicago Press, 1978.

Eliade, Mircea. *Images and Symbols.* Kansas City: Sheed Andrews and McMeel, 1961.

Eliade, Mircea. *The Myth of the Eternal Return.* London: Routledge & Kegan Paul, 1955.

Ferris, T. *The Red Limit.* Toronto: Bantam Books, 1979.

Feynman, Richard. *The Character of Physical Law.* Boston: The MIT Press, 1967.

Frankfort, H.; Frankford, H. A.; Wilson, John A.; Jacobsen, Thorkild; and Irwin, William A. *Before Philosophy: The Intellectual Adventure of Ancient Man.* Chicago: University of Chicago Press, 1946.

Fraser, J. T. *The Genesis and Evolution of Time.* Amherst: University of Massachusetts Press, 1982.

Fraser, J. T., ed. *The Voices of Time.* New York: George Braziller, 1966.

Frey, D. *Gotic und Renaissance als Grundlagen der Modernen Weltanschaung.* Filsberg: Augsburg, 1929.

Frisby, John P. *Seeing.* Oxford: Oxford University Press, 1980.

Frisch, K. V. *The Dance Language and Orientation of Bees.* Cambridge, Mass.: Harvard University Press, 1967.

Gablik, Suzi. *Progress in Art.* New York: Rizzoli International Publications, 1976.

Gardner, H. *Frames of Mind.* New York: Basic Books, 1983.

Gardner, M. *The Relativity Explosion.* New York: Random House, 1976.

Geroch, Robert. *General Relativity from A to B.* Chicago: University of Chicago Press, 1978.

Geymonat, Ludovico. *Galileo Galilei.* Translated by S. Drake. New York: Mc-Graw-Hill, 1957.

Giedion, S. *Space, Time and Architecture.* Cambridge, Mass.: Harvard University Press, 1962.

Gillispie, Charles Coulston. *The Edge of Objectivity.* Princeton, N.J.: Princeton University Press, 1960.

Golding, John. *Cubism,* 2nd ed. London: Faber and Faber, 1968.

Gombrich, E. H. *Art and Illusion.* London: Phaidon Press, 1959.

Gombrich, E. H. *The Story of Art,* 12th ed. London: Phaidon Press, 1972.

Grant, Edward, ed. *A Source Book in Medieval Science.* Cambridge, Mass.: Harvard University Press, 1975.

Graves, Robert, and Patai, Raphael. *Hebrew Myths: The Book of Genesis.* New York: McGraw-Hill, 1963.

Gregory, R. L. *Concepts and Mechanisms of Perception.* London: Gerald Duckworth, 1974.

Gregory, R. L. *The Eye and the Brain.* London: Weidenfeld & Nicolson, 1972.

Gregory, R. L. *The Intelligent Eye.* New York: McGraw-Hill, 1970.

Gregory, R. L., and Gombrich, E. H. *Illusion in Nature and Art.* London: Gerald Duckworth, 1973.

Groenewegen, H. A.; Frankfort, H.; and Ashmole, Bernard. *Art of the Ancient World.* New York: Harry N. Abrams, 1972.

Grout, D. J. *A History of Western Music.* New York: W. W. Norton, 1973.

Hadamard, J. S. *An Essay on the Psychology of Invention in the Mathematical Field.* New York: Dover Publications, 1954.

Hafner, E. M. *The New Reality in Art and Science. Comparative Studies in Society and History.* Vol. 11, p. 385, 1969.

Hamner, K. C.; Finnegan, J. C.; Strohi, G. S.; and Hoshizaki, T. "The Biological Clock at the South Pole," *Nature, 195,* (1962): p. 476.

Hardy, G. H. *A Mathematician's Apology.* Cambridge: Cambridge University Press, 1967.

Harrison, Edward R. *Cosmology: The Science of the Universe.* Cambridge: Cambridge University Press, 1981.

Hartmann, W. H. *Astronomy, The Cosmic Journey.* Belmont, Mass.: Wadsworth Publishing, 1985.

Hartt, Frederick. *Art: A History of Painting, Sculpture, Architecture,* vols 1 and 2. New York: Harry N. Abrams, 1976.

Hartt, Frederick. *Donatello.* New York: Harry N. Abrams, 1972.

Hawking, S. W. "The Quantum Mechanics of Black Holes." *Scientific American,* January (1977): p. 236.

Heath, Sir Thomas. *Euclid.* New York: Dover Publications, 1956.

Henderson, L. D. "A New Facet of Cubism." *The Art Quarterly (1971):* Vol. 34 p. 410.

Herbert, Nick. *Quantum Reality.* New York: Anchor Press, 1985.

Hoffmann, Banesh. *The Strange Story of the Quantum,* 2nd ed. New York: Dover Publications, 1959.

Hoffmann, Banesh, with collaboration of Helen Dukas. *Albert Einstein, Creator and Rebel.* New York: Viking Press, 1972.

Holton, Gerald, and Elkana, Yehuda, eds. *Albert Einstein: Historical and Cultural Perspectives.* Princeton, N.J.: Princeton University Press, 1982.

Hoppin, R. H. *Medieval Music.* New York: W. W. Norton, 1978.

Howe, R. H. *Music Through Sources and Documents.* Englewood Cliffs, N.J.: Prentice-Hall, 1978.

Huang, K., ed. Physics and Our World. A Symposium in Honor of Victor F. Weisskopf. New York: American Institute of Physics, 1976.

Hubel, D. H. Exploration of the primary visual cortex 1955–78. Nature, *299,* 1982, p. 515.

Hughes, Robert. *The Shock of the New.* New York: Alfred A. Knopf, 1981.

Hull, L. W. H. *History and Philosophy of Science.* London: Longmans, Green, 1959.

Huxley, A. *After Many a Summer Dies the Swan.* London: Chatto & Windus, 1950.

Ivins, William Jr. *Art and Geometry.* New York: Dover Publications, 1964.

Jacob, Francois. *The Possible and the Actual,* New York: Pantheon Books, 1982.

Jacob, Francois. *The Logic of Life.* New York: Pantheon Books, 1974.

Jaki, Stanley L. *Science and Creation.* New York: Science History Publications, 1974.

Jammer, M. *Concepts of Space.* New York: Harper & Row, 1960.

Janson, H. W. *The History of Art.* Englewood Cliffs, N.J.: Prentice-Hall, and New York: Harry N. Abrams, 1963.

Jastrow, R., and Thompson, M. H. *Astronomy.* New York: John Wiley & Sons, 1984.

Jerison, Harry. *Evolution of the Brain and Intelligence.* New York: Academic Press, 1973.

Jerison, Harry. "Paleoneurology and the Evolution of the Mind." *Scientific American,* January (1976): p. 90.

Jerison, Harry. CA Book Review, Current Anthropology (1975): *16.* p. 416.

Julesz, B. *Foundation of cyclopean perception.* Chicago: Univ. of Chicago Press, 1971.

Kaufmann, William J. III. *Black Holes and Warped Spacetime.* Toronto: Bantam Books, 1980.

Kaufmann, William J. III. *Universe.* San Francisco: W. H. Freeman, 1985.

Keeton, William T. *Biological Science.* New York: W. W. Norton, 1967.

Kline, Morris. *Mathematics in Western Culture.* London: George Allen and Unwin, 1954.

Kohler, Ivo. *Psychological Issues,* (1964): vol. 3, part 4, p. 19.

Koyre, A. *Galilean Studies,* New York: Humanities Press, 1978. Translated by J. Mepham.

Kramer, E. E. *The Nature and Growth of Modern Mathematics.* Princeton, N.J.: Princeton University Press, 1981.

Kramer, G. "Experiments on Bird Orientation." *Ibis 94.* (1952): p. 265.

Kuh, Katharine. *The Artist's Voice.* New York: Harper & Row, 1960.

Lang, P. H. *Music in Western Civilization.* New York: W. W. Norton, 1941.

Langer, Susanne K. *Feeling and Form.* New York: Charles Scribner's Sons, 1953.

Langer, Susanne K. *Philosophy in a New Key,* 3rd ed. Cambridge, Mass.: Harvard University Press, 1957.

Laporte, P. M. "Cubism and Relativity (With a Letter of Albert Einstein)." *Art Journal* 25. (1947): p. 246.

Lettvin, J. Y., et al. *What the Frog's Eye Tells the Frog's Brain.* Proc. Inst. Radio Engineers, 1959. vol, 47, p. 1940.

Levi-Strauss, Claude. *Myth and Meaning.* Toronto: University of Toronto Press, 1978.

Levi-Strauss, Claude. *The Raw and the Cooked.* New York: Harper & Row, 1969.

Livingston, D. M. *The Master of Light.* New York: Charles Scribner's Sons, 1973.

Luce, G. G. *Biological Rhythms in Human and Animal Physiology.* New York: Dover Publications, 1971.

Lynton, Norbert. *The Story of Modern Art.* Ithaca: Cornell Univ. Press, 1980.

Machlis, Joseph. *The Enjoyment of Music,* 5th ed. W. W. Norton, New York, 1984.

Margenaui, H. *The Nature of Physical Reality. New York: McGraw-Hill, 1950.*

Marr, David. *Vision.* San Francisco: W. H. Freeman, 1982.

Marshack, A. *The Roots of Civilization.* New York: McGraw-Hill, 1972.

Matthews, G. T. V. *Bird Navigation,* 2nd ed. Cambridge: Cambridge University Press, 1968.

McCormmack, R. *Night Thoughts of a Classical Physicist.* Cambridge, Mass.: Harvard Univ. Press, 1982.

Miller, A. I. *Albert Einstein's Special Theory of Relativity.* Reading: Addison-Wesley, 1981.

Millikan, R. The Physical Review Vol. 7, p. 18, 1916 (quoted in PAIS).

Misner, C.; Thorne, K.; and Wheeler, J. *Gravitation.* San Francisco: W. H. Freeman, 1973.

Monod, Jacques. *Chance and Necessity.* New York: Alfred A. Knopf, 1971.

Moore-Ede, M. C.; Sulzman, F. M.; and Fuller, C. A. *The Clocks That Time Us.* Cambridge, Mass.: Harvard University Press, 1982.

Morris, Richard. *Time's Arrows.* New York: Simon & Schuster, 1983.

Mumford, Lewis. *Technics and Civilization.* New York: Harcourt, Brace, 1934.

Needham, J.; Ling, W.; and de Solla Price, D. J. *Heavenly Clockwork.* Cambridge: Cambridge University Press, 1960.

Newton, Sir Isaac. *Principia.* Translated by F. Cajori. Berkeley: University of California Press, 1962.

Onians, John. *Art and Thought in the Hellenistic Age.* London: Thames & Hudson, 1979.

Osborne, Harold. *Abstraction and Artifice in Twentieth Century Art.* Oxford: Oxford University Press, 1979.

Pagels, H. *The Cosmic Code.* New York: Simon & Schuster, 1982.

Pagels, H. *Perfect Symmetry.* New York: Simon & Schuster, 1985.

Pais, Abraham. *'Subtle is the Lord. . . ': The Science and the Life of Albert Einstein.* Oxford: Clarendon Press, 1982.

Park, Davis. *The Image of Eternity.* Amherst: University of Massachusetts Press, 1980.

Pascal, B. *The Pensées.* Translated by J. M. Cohen. Harmondsworth: Penguin Books, 1961.

Pengelley, Eric T., and Asmundson, Sally J. "Annual Biological Clock." *Scientific American,* April 1971, p. 72.

Penrose, R. "Black Holes." *Scientific American,* May, 1972, p. 38.

Peyser, Joan. *The New Music.* New York: Delacorte Press, 1971.

Pfeiffer, J. *The Creative Explosion.* New York: Harper & Row, 1982.

Philosophical Transactions of the Royal Society of London. *The Psychology of Vision* (1980): *290.* p. 1.

Piaget, Jean. *The Child and Reality.* New York: Grossman Publishers, 1973.

Piaget, Jean. *Genetic Epistemology.* New York: Columbia University Press, 1970.

Piaget, Jean. *The Principles of Genetic Epistemology.* New York: Basic Books, 1972.

Piaget, Jean. *Psychology and Epistemology.* New York: Grossman Publishers, 1971.

Pilkington, J. G., trans. *The Confessions of St. Augustine.* New York: Liveright, 1942.

Pirenne, M. H. *Optics, Painting, Photography.* Cambridge: Cambridge University Press, 1970.

Popper, Karl R. *Objective Knowledge.* Oxford: Clarendon Press, 1979.

Renoir, Jean. *Renoir, My Father.* Boston: Little, Brown, 1962.

Richardson, John A. *Modern Art and Scientific Thought.* Urbana: University of Illinois Press, 1971.

Rock, I., and Harris, C. S. *Scientific American,* May 1967, p. 96.

Rosen, Charles. *Arnold Schoenberg*. New York: The Viking Press, 1975.

Rosen, Charles. *The Classical Style: Haydn, Mozart, Beethoven.* New York: The Viking Press, 1971.

Rosen, J. *Symmetry Discovered.* New York: Cambridge University Press, 1975.

Rosenblatt, S. trans. *Saadia Gaon.* New Haven, Conn.: Yale University Press, 1948 (quoted in DAVIS and HERSH).

Russell, Bertrand. *A History of Western Philosophy.* New York: Simon & Schuster, 1945.

Russell, John. *The Meanings of Modern Art.* New York: Harper & Row, 1981.

Sambursky, S. *The Physical World of Late Antiquity.* London: Routledge & Kegan Paul, 1962.

Sambursky, S. *The Physical World of the Greeks.* London: Routledge & Kegan Paul, 1956.

Sapir, E. *Selected Writings in Language, Culture and Personality.* Berkeley: University of California Press, 1958.

Schmidt, George. *Form in Art and Nature.* Basel: Basilius Press, 1959.

Schoenberg, Arnold. *Style and Idea.* London: Faber & Faber, 1975.

Seelig, C. *Albert Einstein.* London: Staples Press, 1956.

Segall, M. H.; Campbell, D. T.; and Herskovits, M. J. "Cultural Differences in the Perception of Geometric Illusions." *Science (1963): 139.* p. 769.

Simpson, George Gaylord. *Biology and Man.* New York: Harcourt Brace Jovanovich, 1969.

Simpson, George Gaylord. *The Meaning of Evolution.* New Haven, Conn.: Yale University Press, 1967.

Smith, B., ed. *Concerning Contemporary Art.* Oxford: Clarendon Press, 1967.

Smith, J. Maynard. *The Theory of Evolution.* Harmondsworth: Penguin Books, 1976.

Sperry, R. W. *Scientific American,* May 1956, p. 48.

Steinberg, L. *Other Criteria.* New York: Oxford University Press, 1972.

Steiner, George. *Language and Silence.* New York: Atheneum, 1967.

Stravinsky, Igor. *Chronicle of My Life.* London: Victor Gollancz, 1936.

Stravinsky, Igor. *Poetics of Music.* Cambridge, Mass.: Harvard University Press, 1977.

Stuckenschmidt, H. H. *Twentieth Century Music.* New York: McGraw-Hill, 1969.

Sullivan, W. *Black Holes.* New York: Warner Books, 1980.

Thomas, L. *The Medusa and the Snail.* New York: The Viking Press, 1970.

Toulmin, Stephen, and Goodfield, June. *The Discovery of Time.* Harmondsworth: Penguin Books, 1967.

Trefil, J. *From Atoms to Quarks.* New York: Charles Scribner's Sons, 1980.

Trefil, J. *The Moment of Creation.* New York: Collier Books, 1983.

Uttal, W. R. *The Psychobiology of Mind.* Hillsdale: Lawrence Eribaum Publishers, 1978.

Vallier, Dora. *Abstract Art.* New York: The Orion Press, 1970.

Van der Waerden, B. L. "The Great Year in Greek, Persian and Hindu Astronomy." *Archive for History of Exact Sciences, 18.* (1978): p. 359.

Vasari, G. *The Lives of Painters, Sculptors and Architects.* Edited by William Gaunt. London: J. M. Dent & Sons, 1963.

Von Senden, M. *Space and Sight.* Glencoe, Ill.: The Free Press, 1960.

Waddington, C. H. *Behind Appearance.* Cambridge, Mass.: The MIT Press, 1970.

Waldrop, M. M. "Computer Vision." *Science,* vol 224, (1984): p. 1225.

Warrington, John, trans. and ed. *Plato's 'Timaeus.'* London: J. M. Dent & Sons, 1960.

Wax, M. *American Journal of Sociology 65.* (1960): p. 449.

Webb, B. *Biological Rhythms, Sleep and Performance.* New York: John Wiley & Sons, 1982.

Wechsler, Judith, ed. *On Aesthetics in Science.* Cambridge, Mass.: The MIT Press, 1981.

Weinberg, Steven. *The First Three Minutes.* New York: Basic Books, 1977.

Weiner, C., ed. *History of Twentieth Century Physics.* New York: Academic Press, 1977.

White, John. *The Birth and Rebirth of Pictorial Space.* Boston: Boston Book and Art Shop, 1967.

Whitrow, G. J. *The Natural Philosophy of Time,* 2nd ed. Oxford: Clarendon Press, 1980.

Whorf, B. L. *Language, Thought and Reality.* New York: John Wiley & Sons, 1959.

Wiesel, T. N. Postnatal development of the visual cortex and the influence of environment. Nature, *299,* p. 583, 1982.

Wigner, E. *Symmetries and Reflections.* Bloomington: Indiana University Press, 1967.

Wilson, Richard Albert. *The Birth of Language.* Toronto: University of Toronto Press, 1980.

Young, J. Z. "Brains and Worlds: The Cerebral Cosmologies." *Journal of Experimental Biology, 61.* (1974): p. 5.

Young, J. Z. *Programs of the Brain.* Oxford: Oxford University Press, 1978.

Zammatio, Carlo; Marinoni, Augusto; and Brizio, Anna Maria. *Leonardo the Scientist.* New York: McGraw-Hill, 1980.

Zarlino, Gioseffo. *The Art of Counterpoint.* Translated by Guy A. Marco and Claude V. Palisca. New Haven, Conn.: Yale University Press, 1968.

Zuckerkandl, Victor. *Sound and Symbol.* Princeton, N.J.: Princeton University Press, 1969.

Notes, Sources, Reading Suggestions

(The names in small caps (MONOD) refer to the entries in the bibliography)

CHAPTER II

P. 13.

Chapters II and III are largely about biological evolution. The basic ideas describing this process were developed already in the nineteenth century mainly (but not exclusively) by Charles Darwin, Alfred Wallace, and Gregor Mendel. The important twentieth-century discoveries in biology—including the discovery of the molecular structure of the genetic material by Francis Crick and James Watson—all turned out to be compatible with evolutionary theory. Among the many books dealing with the modern form of this theory are SIMPSON (1967), MONOD, SMITH, and JACOB (1974).

P. 14.

Short summaries of the properties of a large number of biological clocks are given in BRADY (1982) and BRADY (1979), in the conference proceedings at Cold Spring Harbor COLD SPRING HARBOR SYMPOSIA and in CLAUDSLEY-THOMPSON. The circadian clocks are discussed in MOORE-EDE, SULZMAN and FULLER. For their effects on normal human life, see LUCE and WEBB.

P. 16.

A nontechnical discussion of circannual clocks can be found in PENGELLEY and ASMUNDSON and also in CLAUDSLEY-THOMPSON. The South Pole experiments were reported in *Nature* by HAMNER, FINNEGAN, STROHI and HOSHIZAKI.

P. 18.

The experiments of W. W. Gardner and H. A. Allard are discussed in many comprehensive introductory biology texts. See, for example, KEETON.

P. 19.

Details about the nature of photoperiodic clocks can be found in BUNNING.

P. 20.

Karl von Frisch's observations on the biological clocks and other properties of bees are described in FRISCH.

P. 21.

The experiments clarifying some important environmental clues of bird migration were first reported in 1950 in Germany. An English summary of these can be found in KRAMER (quoted in BRADY, 1982). More details and further examples on animal navigation are presented in MATTHEWS and in CLAUDSLEY-THOMPSON.

P. 24.

The quotation by R. L. GREGORY is from his 1970 book. Some of the more useful nontechnical introductory texts into the problems of seeing are GREGORY (1974), GREGORY (1972), GREGORY (1970), PIRENNE and FRISBY. A more technical treatment of this subject was written by the pioneer of the contemporary theory of seeing, the late DAVID MARR of M.I.T. The Royal Society conference quotation is from PHILOSOPHICAL TRANSACTIONS OF THE ROYAL SOCIETY OF LONDON. Current research in machine vision is summarized in a short report on computer vision by M. MITCHELL WALDROP.

P. 26.

The Gifford lectures of Z. J. Young give a readable account of "models" of the brain YOUNG (1978).

P. 27.

R. W. Sperry's experiments are explained in nontechnical terms in SPERRY. The Innsbruck "upside-down world" experiments are described in KOHLER.

P. 28.

The expression "cosmologies" for the internal world models of organisms was suggested by Z. J. Young in his 1974 paper.

The famous experiments by LETTVIN et al. were originally described in an engineering journal.

P. 30.

The quotation is from the film director JEAN RENOIR's biography of his painter-father.

CHAPTER III

P. 31.

The first quotation is from POPPER.

A detailed description of Jerison's work and ideas are given in JERISON (1973); a more popular exposition is JERISON (1976).

P. 32.

The quotation is from JERISON (1973).

P. 35.

The quotation is from JERISON (1973).

P. 36.

"Complex reflex machine" is from JERISON (1975).

P. 37.

All quotations by F. JACOB are taken from his 1982 book.

P. 40.

M. VON SENDEN's work is described in his 1960 book and also in GREGORY (1974).

The experiments with kittens are described in BLAKEMORE and COOPER. On the problem of visual training see also the Nobel acceptance talk by WIESEL.

Some of Piaget's ideas about the evolution of the innate and acquired knowledge of the external world are summarized in his works: PIAGET (1973), PIAGET (1972), PIAGET (1971), and PIAGET (1970).

P. 43.

K. J. W. Craik's ideas are described in his 1952 book.

P. 44.

Jerison's speculations concerning language are presented in JERISON (1976) and JERISON (1973).

P. 46.

The quotation is from JERISON (1973).

The footnote is from SAPIR.

P. 49.

All quotations by S. Langer are from LANGER (1957). In this book, Langer actually used the word "sign" for what we call here a "signal." In LANGER (1953), however, which she regarded in effect as volume II to LANGER (1957), Langer expressed preference for the use of these words as given here.

P. 50.

The quotation is from SIMPSON (1969).

P. 51.

The quotation is from BOULDING.

P. 55.

E. Sapir's remarks are from his 1958 book.

B. Whorf's ideas are summarized in WHORF (1959).

The Wilson quotation is from WILSON (1980).

CHAPTER IV

P. 57.

MARSHACK contains interesting data concerning the evolution of symbolism. PFEIFFER presents some interesting speculations about the origins of art.

All quotations by H. Frankfort are from FRANKFORT et al.

P. 61.

M. ELIADE's remark is from his 1961 book.

In the 1920s, the Austrian art historian DAGOBERT FREY studied the links between the notions of time and space and the art forms in different civilizations.

A general survey of ancient art containing further literature is GROENEWEGEN.

P. 66.

The words of M. ELIADE are from his 1978 book.

The quotation by M. ELIADE is from his 1955 book.

The Babylonian Great Year is described, for example, in CAMPBELL. For more detail about the recurring cycles in various civilizations, see VAN DER WAERDEN.

P. 69.

Max Weber's conjecture is discussed in WAX. For a readable historical analysis of the Genesis legend, see, for example, GRAVES and PATAI.

P. 70.

Plato's cosmology is described mainly in his "Timaeus." It is here that he defines time as the "moving likeness of eternity" and says that the very process of time is caused by the motion of heavenly bodies WARRINGTON. Timaeus is discussed in great detail in CORNFORD (1948).

P. 71.

I learned about the origins of Ecclesiastes from Professor A. Kasher of Tel-Aviv University who summarized them in a private communication to my colleague Dr. M. Schlesinger. My thanks to both of them.

P. 72.

The opinion of BERTRAND RUSSELL is taken from his 1945 book.

P. 76.

For Plato's remarks about astronomy, see CORNFORD (1968) or BLOOM.

P. 77.

Delbrück's remark is from his contribution in HUANG.

P. 78.

Galen's opinion about beauty is quoted in DE LA CROIX and TANSEY.

P. 81.

Euclid's foundation of geometry is available in English translation in HEATH.

Concerning the compatibility of touch and vision, see ROCK and HARRIS.

The idea that the ancient Greeks possessed a more developed sense of touch is an old one. See, for example, IVINS.

P. 82.

Plato's disparaging remarks about paintings are in his *Republic*. See CORNFORD (1968) and BLOOM.

P. 83.

The works of Greek mathematicians and scientists are described, for example, in BOYER, HULL, SAMBURSKY (1962), and SAMBURSKY (1956). The characteristic ideas of the Hellenistic era are discussed in ONIANS.

P. 87.

The quote by LEWIS MUMFORD is from his 1934 book.

CHAPTER V

P. 88.

The quotation by I. STRAVINSKY is from his 1936 autobiography.

P. 90.

The text quoted by H. BUTTERFIELD is from his 1957 book.

P. 91.

Newton's definition of absolute time is from NEWTON. It is instructive to read the whole paragraph:

"Absolute, true and mathematical time, of itself, and from its own nature, flows equably without relation to anything external, and by another name is called duration: relative, apparent, and common time, is some sensible and external (whether accurate or unequable) measure of duration by the means of motion, which is commonly used instead of true time; such as an hour, a day, a month, a year."

P. 92.

For the Mueller-Lyell and other types of illusion, see, for example, GREGORY (1970).

P. 96.

The Chinese clocks are described in NEEDHAM, LING and DE SOLLA PRICE.

P. 97.

The words by D. J. de Solla Price are from his 1961 book.

About candles as time-measuring instruments, see BEDINI.

P. 98.

PIAGET's remarks are from his 1971 book.

P. 99.

There are several current books offering more detailed histories of the evolution of the concepts of time from various viewpoints. Among them are WHITROW, PARK, JAKI, MORRIS, FRASER (1982), FRASER (1966) and TOULMIN and GOODFIELD. Also, there are several interesting articles in the series "The Study of Time," published regularly by Springer-Verlag.

St. Augustine's ideas about time are mostly in his *Confessions* PILKINGTON.

A monumental work by M. CLAGETT containing translations of the most important medieval works on mechanics does not have the name of St. Augustine in the index. In another rich collection of medieval scientific texts by E. GRANT, St. Augustine's time ideas are mentioned just once and then in a context that is irrelevant to our problem. These facts illustrate the extent to which St. Augustine's ideas were disregarded in the science of the Middle Ages.

Soon after Galileo, Descartes also found the law of free-fall by expressing the distance covered as a function of time (KOYRE).

P. 100.

Concerning Galileo's experimental work, see DRAKE.

For more details about the history of polyphony, see, for example, CALDWELL, HOPPIN, and APEL.

P. 103.

The words about Perotin are from DITTNER. On the performance of polyphonic music see BUKOFZER.

P. 105.

Again, the words are from DITTNER.

The book by M. CLAGETT is the one quoted earlier.

P. 107.

The quotations by V. ZUCKERKANDL are from his 1969 book.

P. 109.

The words by Pope John XXI are quoted, for example, in HOWE.

CHAPTER VI

P. 112.

The quotation by A. HUXLEY is from his 1950 book.

About visual perception, see, for example, GREGORY (1974), GREGORY (1972), GREGORY (1970), PIRENNE, FRISBY and MARR, about the complexity of binocular vision JULESZ, while vision and art are discussed in GOMBRICH (1959) and ARNHEIM.

P. 116.

The quotation by G. VASARI is from a 1963 edition of his book.

P. 117.

There are many excellent art histories relating the discovery of linear perspective and its consequences. See, for example, DE LA CROIX and TANSEY, GOMBRICH (1972), JANSON, HARTT (1976), GABLIK, and EDGERTON.

P. 118.

The principles of perspective are discussed in BARTSCHI.

P. 119.

The quotations by F. HARTT are from his 1972 book.

P. 120.

The words by Janson are from his 1963 book.

P. 121.

The quotation is from WHITE.

P. 123.

The linear perspective controversy is discussed in EDGERTON, where further references are also given. For non-Western perception of perspective clues, see, for example, V. B. Deregowski's contribution in GREGORY and GOMBRICH or SEGALL et al.

P. 124.

The quotation by Leonardo is taken from ZAMMATIO, MARINONI, and BRIZIO.

P. 125.

The quotations are from WHITE.

P. 127.

Regarding the relation of linear perspective to projective geometry, see, for example, KLINE.

P. 128.
The quotation by Descartes was taken from GILLISPIE.

P. 129.
NEWTON's words are from Cajori's translation.
The history of the ideas of space is presented in JAMMER.

P. 133.
The text by PASCAL was taken from Cohen's translation.

P. 136.
For NEWTON's laws of mechanics, see Cajori's translation.
The words about the center of the world are from NEWTON.

P. 137.
The Galileo translation is from GEYMONAT.

P. 144.
Michelson's statement is quoted in LIVINGSTON.

CHAPTER VII

P. 145.
The quotation is from KUH.

P. 146.
There are a large number of nontechnical books about the theories of relativity, of the space and time of the universe, of black holes, etc. A few are listed here: BORN, BONDI, GARDNER, ASIMOV, BERGMANN, DAVIES (1978), DAVIES (1977), GEROCH, HARRISON, KAUFMANN and SULLIVAN. A lot of details about these theories, most of which I could not discuss in such a short space, can be found in many of these books.

Concerning the history of the creation of the special and general theories, see the excellent scientific biography of Einstein by A. PAIS. Parts of this book are technical, but even if one must omit those, the book still remains highly interesting. A more detailed technical history of the special theory is presented in MILLER. For interesting aspects of Einstein's creative life, see BERNSTEIN or HOFFMANN (1972). On the influence of relativity on other areas of thinking, see HOLTON and ELKANA.

P. 156.
PIAGET's problem is related in his 1970 book.

P. 160.
Michelson's words are from LIVINGSTON.
The psychological background underlying the insistence of the reality of the ether is depicted in a novel by McCormmack.

P. 167.
A short history of the equation $E = mc^2$ is given in PAIS; a more detailed one in MILLER. A nontechnical discussion of gravitation and space-time is in BERGMAN.

P. 175.
The quotation is from MISNER, THORNE and WHEELER.

P. 176.

The Eddington expedition is described in PAIS.

P. 180.

Black holes and singularities are discussed without mathematics in PENROSE.

HAWKING's ideas of the role of quantum effects in black holes are described in nontechnical terms in his 1977 article.

CHAPTER VIII

P. 182.

The quotation by FEYNMAN is from his 1967 book.

A careful and easily readable introduction into the history of quantum mechanics was written by BANESH HOFFMANN (1959). More detailed histories are presented in PAIS and WEINER. The Planck-Einstein discoveries are treated in WEINER by M. Klein.

P. 184.

The quotations by Einstein are from ARONS and PEPPARD.

P. 185.

The Planck quotation was taken from SEELIG.

The remarks by MILLIKAN are from his technical paper.

The Compton effect is described, for example, in HOFFMANN (1959) and in WEINER.

P. 186.

For the history of atomic phenomena, see J. L. Heilborn's paper in WEINER.

P. 188.

The initial controversy between the founders of quantum mechanics is described by A. I. Miller in WECHSLER.

P. 190.

For nontechnical descriptions of the uncertainty principle and related problems, see, for example, FEYNMAN, DAVIES (1980), PAGELS (1982), TREFIL (1980), and TREFIL (1983).

P. 199.

A most recent and quite spectacular illustration of the impossibility of describing the motion of elementary particles in space and time in any familiar intuitive way was provided by a series of experiments of A. Aspect and his co-workers during the first half of the 1980s. The idea of these experiments came from a famous theoretical analysis written in 1935 by Albert Einstein, Boris Podolsky and Nathan Rosen at the Princeton Institute of Advanced Studies. Subsequently, in 1964, John Bell, then at the University of Wisconsin, devised a testable consequence of the above-mentioned analysis. The phenomenon, roughly speaking, is the following: Imagine two particles which interacted with each other at some time in the past and are now far from each other. Far enough that no possible further interaction can take place between them. In spite of this, quantum mechanics predicts that in certain cases a measurement carried out on one of the particles will instantaneously change the state of the other no matter how far the particles are from each other. While this prediction sounds quite absurd intuitively, this is the prediction which has been nevertheless verified by the Paris experiments. For (non-technical) details and for further references, see the papers by B. d'Espagnat and D. Mermin, or the books by PAGELS (1982) or HERBERT.

P. 201.

A more detailed philosophical scrutiny of the Pauli principle is given in MARGENAU.

P. 203.

Dirac's theory of the vacuum is described in detail in DAVIES (1980), PAGELS (1982), TREFIL (1980), and DAVIES (1984).

P. 210.

The growing inadequacy of ordinary language in some areas of contemporary life and culture is discussed in STEINER.

Interesting ideas about mathematics in general are discussed in DAVIS and HERSH and in E. E. KRAMER.

P. 212.

The quotations by WIGNER are taken from his 1967 book.

The first successful theory of strongly interacting particles was suggested in 1961 by M. Gell-Mann of the California Institute of Technology and the Israeli physicist Y. Neeman, then at the Imperial College in London. Their work made use of abstract group theory without reference to properties in space and time. For nontechnical expositions of the underlying ideas, see CHEW et al. or DAVIES (1979).

CHAPTER IX

P. 214.

The quotation by Poincare is taken from CHANDRASEKHAR.

Some of the books that describe modern art are RUSSELL, GABLIK, LYNTON, HUGHES, and VALLIER.

P. 217.

The quotation is from DE LA CROIX and TANSEY.

P. 218.

The quotation by GOLDING is from the 1968 edition of his book.

P. 219.

According to the now classic experiments of D. Hubel and T. Wiesel, we seem to have cells in our brain that detect well-defined visual features such as edges and angles. FRISBY describes these effects. For additional information, see UTTAL or for a short review see the Nobel acceptance speech in HUBEL.

P. 223.

The first quotation is taken from OSBORNE.

The quotation by Heron is from his contribution in SMITH.

HARDY's words are from his 1967 book.

P. 227.

The alleged similarities between cubist and relativist space were proposed in GIEDION and LAPORTE. Einstein's rebuttal is in LAPORTE. A more thorough recent study of the problem is given in HENDERSON.

P. 228.

STEINBERG's paper is in his collected papers.

The catalogue of the Basel exhibition is available in English (SCHMIDT).

P. 229.

WADDINGTON is an interesting study of the relations between painting and the natural sciences in this century by the distinguished geneticist C. H. Waddington. For other aspects of this problem, see RICHARDSON.

HADAMARD's results are presented in his 1954 book.

Visual and mathematical abilities are also discussed in GARDNER.

P. 230.

The evolution of twentieth century literature along the lines of steering away from continuity and causal predictability by the "shattering of the narrative line" was described recently by the critic ANNIE DILLARD.

The remarks of Lewis Thomas and C. Levi-Strauss are from THOMAS, LEVI-STRAUSS (1969), and LEVI-STRAUSS (1978) respectively.

P. 231.

For a general history of music, see, for example, LANG, EINSTEIN, ABRAHAM, and GROUT. The basic ideas of music theory are explained in COPLAND and MACHLIS.

ZARLINO's words are from the 1968 translation of his book.

The quotation by CHARLES ROSEN is from his 1971 book.

P. 233.

The properties of musical motion are described, for example, in LANGER (1953) or in ZUCKERKANDL.

STRAVINSKY's words are from his 1977 book.

P. 234.

About the evolution of homophony, see any standard music history, such as LANG or EINSTEIN.

P. 237.

Atonality and its further developments are discussed in C. ROSEN (1975). More technical ideas can be found in SCHOENBERG. Modern music in general is discussed, for example, in STUCKENSCHMIDT, PEYSER, and COPLAND (1968).

P. 239.

The similarities between the layman's view of modern art and modern science are discussed in HAFNER. Another opinion by the science historian T. Kuhn is printed at the end of this paper.

CHAPTER X

P. 240.

Saadia Gaon is considered the most important Jewish philosopher in the tenth century. The quotation is from ROSENBLATT.

More detailed explanations of most of the material presented here can be found in PAGELS (1985), TREFIL (1983), and DAVIES (1979). Among the many current nontechnical books on cosmology and astronomy, I mention a few only: HARRISON, WEINBERG, FERRIS, KAUFMANN, JASTROW and THOMPSON, and HARTMANN.

P. 245.

The current form of the inflationary universe is described in nontechnical terms in a *Scientific American* paper by A. GUTH and P. J. STEINHARDT. See also the books by DAVIES (1984), PAGELS (1985) and TREFIL (1983). The quotation about the universe as a vacuum fluctuation is from TRYON.

P. 250.

The idea of symmetry is treated in a nontechnical manner in J. ROSEN.

Acknowledgements

In covering a variety of subjects I have accumulated a great many debts. It is a great pleasure to acknowledge at least some of them.

Yoram Avidor, John Dell, Amos Harpaz, Luc Longtin, and Byron Rourke read parts of the manuscript and gave valuable advice. Rick Buzzeo helped with some drawings, and Allison Price with literature search. I am also very thankful to Doreen Pullen and Petrona Parungo for typing the earlier drafts of the manuscript and to Barbara Ridsdale for typing, correcting and organizing the final product.

From its conception, this book was also a family enterprise. My wife taught me many things about the arts, and she also read the manuscript with a critical eye. My son Michael and my daughter-in-law Elisabeth were also generous with their time and advice. I learned a lot about music and music history from my brother

Edwin and was able to discuss my problems in biology and in almost everything else with my brother Alfred.

The most important contribution, however, was made by my daughter Annie. She helped me to focus and sharpen my arguments, and it is mainly due to her hard work that the text now largely expresses what I wanted to say.

Index